STEEL ISN'T HARD (TO LEARN)

STEEL IMAGE

SHANE TURCOTT

STEEL IMAGE

Steel Image
7 Innovation Drive, Suite 155
Dundas, ON
L9H 7H9 Canada

www.steelimage.com

Design by Catharine Barker,
National Graphics Toronto,
ON Canada

Printed and bound in Canada

Title: Steel Isn't Hard
(To Learn)
ISBN 978-1-7771576-1-6

STEEL ISN'T HARD (TO LEARN)

STEEL IMAGE

SHANE TURCOTT

TABLE OF CONTENTS

Dedicated to my parents,

Bob and Gail Turcott

Because you fostered curiosity, hard work and the pursuit of useful knowledge.

The world we know today
was built upon the
intelligent use of steel.

1.0 INTRODUCTION

Steel is beautiful. Not because of its appearance but rather for what it has allowed mankind to build. From the world's tallest buildings, massive bridges and monstrous ships, to the everyday things such as cars, appliances, kitchen knives and nails. The more humanity has learned about steel, the more incredible and common our creations have become.

The ability to design, manufacture and repair things requires a practical understanding of the materials being used. Since steel is the most commonly used metal in the world, those striving to become capable engineers, welders, machinists or maintenance personnel will greatly benefit from building a practical foundation in steel. If one masters steel, other metals will become easier to learn.

There are many different types of steel. Each steel grade is designed to balance several attributes such as strength, cost, durability, corrosion resistance and the ease with which it can be manufactured into things. As some traits are optimized, others may become diminished. For example, some additives that increase the steel strength can make it difficult to weld – affecting how a part can be both fabricated and repaired. Knowing the basics of steel will help to ensure that the proper steel is used, the optimized manufacturing steps are employed and, when needed, the appropriate repairs can be made.

Steel can be quite strong and hard – these are some of the reasons why we use it. However, steel is not hard to learn and understand. This book will provide a practical introduction to steel. It will focus on the most critical, most pragmatic concepts relevant to its use. These concepts will be demonstrated to show how they work and to provide context of their importance. Topics covered include alloying, steel properties, processing, welding, water corrosion and stainless steels. After reading this book, you will have a useful foundation that will provide context for further study on topics specific to your career, projects and interests.

Although few consider steel an exciting subject before studying it, from the right perspective it can be fascinating, intellectually gratifying and highly useful. For some, steel metallurgy will begin to satisfy their curiosities in comprehending the man-made things surrounding us. Understanding steel's fundamental behaviors allows for the intelligent use of one of the most critical engineering materials ever discovered. For those who can recognize the opportunity in that, I hope you will share in my enthusiasm for steel throughout this book and afterwards in your career.

These bolts are made from different types of steel. They will be used to demonstrate how alloying and processing can adjust their strengths.

2.0 THE KEY CONCEPTS

Most people intuitively understand that the strength of steel depends on what ingredients are included within the steel. Yet these ingredients, referred to as 'alloying', make up only half the equation. Whether steel is strong or weak, ductile or brittle is also based upon a second factor – how these ingredients are arranged within the steel. In more technical terms, steel properties are based upon both its (a) alloying and (b) 'microstructure' – referencing how the atoms are organized within. To design, manufacture and repair steel parts requires understanding the roles of alloying and the microstructures formed by different processing.

2.1 Three Different Bolts

Perhaps the best way to introduce the key concepts of steel is by demonstrating them. **Figure 2.1** displays three steel bolts. Each is made of a different type of steel with different alloying. Although identical in size, the variation in their alloying will affect how each bolt responds to processing and the strengths that can be achieved.

Figure 2.1: These three bolts are made of different steel grades, each with different alloying. They will be used to demonstrate how alloying, microstructure and processing affect important steel properties such as strength.

Their dissimilar colors were due to differences in their coatings.

Table 2.1: Bolt Grade, Cost and Primary Alloying

Bolt Grade	Grade 2	Grade 5	Grade 9
Size	5/8-inch diameter, 3-inch long hex bolts		
Cost per bolt	\$0.45	\$0.60	\$1.50
Major alloying elements (elements affecting steel properties)	0.15% carbon	0.40% carbon	0.40% carbon 0.20% molybdenum 0.45% chromium 0.45% nickel

Table 2.1 lists the costs and primary alloying of the three bolts classified as Grades 2, 5 and 9. We cannot talk about metal, design or manufacturing without first appreciating the need to control costs (so people will actually buy the parts made). While the Grade 2 and 5 bolts were not too dissimilar in price, the addition of pricey molybdenum, chromium and nickel to the Grade 9 bolt notably increased its cost. It is often cost-effective to use a few extra cheap bolts or thicker steel parts rather than to pay for such expensive alloying. Yet in other cases the benefits obtained from these additives are worth every penny. To understand which bolt makes sense for a particular application, we must consider the capabilities, limitations and costs of the steel types from which they are made of – something we will do throughout this book.

2.2 Alloying, Heat Treatment and Strength

Steel is heat treated at glowing red-hot temperatures to establish key properties such as strength. These heat treatments are analogous to the cooking of food – differences in temperature, time and even how the food is left to cool afterward will affect the final product. Each recipe of different ingredients will have its own optimized cooking conditions. The same is true for steel.

The two most common heat treatments applied to steel are (1) normalizing and (2) quenching and tempering. Normalizing is completed by bringing the steel to sufficiently hot temperatures, holding for some time and then

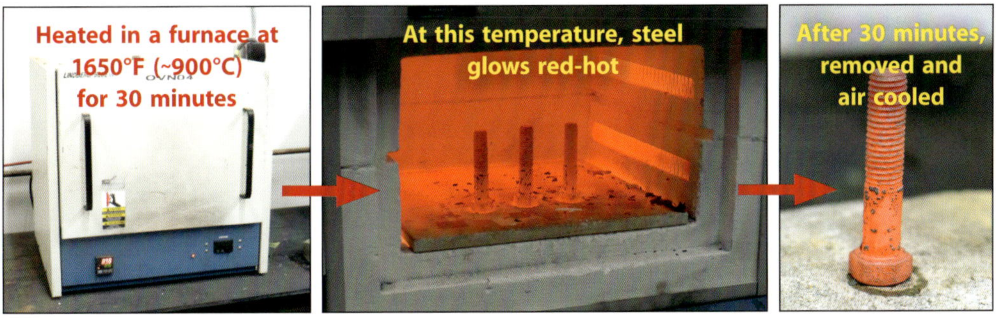

Figure 2.2: The bolts were 'normalized' by heating them up to 1650°F (~900°C) for thirty minutes in a furnace and then removed to air cool.

simply removing it from the furnace to air cool. **Figure 2.2** illustrates how all three bolts were normalized at the red-hot temperature of 1650°F (~900°C) and then air cooled. Even after identical processing, the strength of each bolt was quite different.

Before discussing the outcome of this heat treatment, we should first establish a means to measure steel strength. Most manufactured parts such as bolts, gears, shafts and other engine components will not function properly if they become misshapen. The load that begins to deform or misshape a part (stretching, bending, buckling, etc.) is the upper limit of what a steel part could sustain and still be expected to function. Therefore, we will use the weight threshold that begins to cause irreversible stretching as our primary measure of steel strength for now.

Back to our normalized bolts, **Table 2.2** lists the bolts' strengths after normalizing. To begin to deform and permanently stretch the Grade 2 bolt took 9000 lbs (pounds) of weight. To put that into context, the average weight of a car is just over 4000 lbs. This meant that a single, normalized Grade 2 bolt could hold two cars.

Table 2.2: Bolt Strength after Normalizing Heat Treatment

Bolt Grade	**Grade 2** 0.15% C	**Grade 5** 0.40% C	**Grade 9** 0.40% C + Mo + Cr + Ni
Heat Treatment	**Normalized** (1650°F for 30 minutes, then air cooled)		
Weight Causing Deformation	9000 lbs	15,500 lbs	21,000 lbs

As strong as the Grade 2 bolt might seem, that is not very strong for steel. The Grade 5 bolt could hold the weight of almost four cars – a benefit from its higher carbon content. Carbon is the primary alloying element of steel and the more carbon steel has, the stronger it will be.

The Grade 9 bolt was even stronger. It held an additional SUV in weight above what the Grade 5 bolt could handle. Since both bolts had the same carbon content, clearly the additional strength of the Grade 9 bolt was from its molybdenum, chromium and nickel alloying.

The benefits of carbon and other alloying go beyond this. If enough carbon is present, it can provide a dramatic strength increase when heat treated by the second common method: quenching and tempering. Quenching and tempering starts by heating the steel to similar red-hot temperatures as normalizing. However, instead of air cooling afterward, the steel is submerged in water or oil which rapidly quenches it. In this as-quenched condition, steel could be brittle and unsafe for use. Yet 'tempering' quenched steel by heating it to medium temperatures will restore its ductility (its ability to bend and stretch). **Figure 2.3** shows how new bolts were quenched from 1650°F (~900°C) and then tempered at 900°F (482°C) for one hour. **Table 2.3** lists the bolts' strengths after this heat treatment.

With changing only how they were heat treated, the strengths of two bolts were nearly doubled. It was because the Grade 5 and 9 bolts had enough carbon to benefit from the quenching effect. Whereas the

Figure 2.3: The bolts were 'quenched and tempered' by heating them to 1650°F (~900°C) for thirty minutes, quenching them in water and then tempering at 900°F (482°C).

Table 2.3: Bolt Strength after Quenching and Tempering

Bolt Grade	**Grade 2** 0.15% C	**Grade 5** 0.40% C	**Grade 9** 0.40% C + Mo + Cr + Ni
Heat Treatment	**Quenched and Tempered** (1650°F for 30 minutes, water quenched then 900°F temper)		
Weight Causing Deformation	12,000 lbs	31,000 lbs	38,000 lbs

normalized Grade 5 bolt could hold the weight of nearly four cars, in this quenched and tempered condition, the same bolt grade could now hold eight cars. The quenched and tempered Grade 9 bolt could hold almost ten cars – that is four times stronger than the normalized Grade 2 bolt.

The tempering temperature could also be used to adjust the strengths of the quenched Grade 5 and 9 bolts. **Table 2.4** lists how changing the tempering temperatures altered the bolts' strengths – a behavior useful for tailoring the steel properties for specific applications.

Table 2.4: Effects of Tempering Temperature on Bolt Strength

Bolt Grade	**Grade 2** 0.15% C	**Grade 5** 0.40% C	**Grade 9** 0.40% C + Mo + Cr + Ni
Heat Treatment	**Quenched and Tempered** (1650°F for 30 minutes, water quenched and then tempered)		
Tempering Temperature	**Weight Causing Deformation**		
700°F (371°C)	12,000 lbs	40,500 lbs	47,000 lbs
900°F (482°C)	12,000 lbs	31,000 lbs	38,000 lbs
1100°F (593°C)	11,000 lbs	23,000 lbs	30,000 lbs

While quenching and tempering had dramatically improved the Grade 5 and 9 bolts' strengths compared to their normalized conditions, the

Grade 2 bolt had not sustained the same substantial benefits. The Grade 2 bolt had experienced a strength increase but not of the same magnitude. Neither had adjusting the tempering temperature greatly affected its strength. This was because of its low carbon content and its inability to 'harden' during quenching heat treatments. To properly explain this, we need to discuss what happened to the steel during these heat treatments.

2.3 Microstructures Formed by Heat Treating

The reason why the heat treatments changed the bolts' strengths was that they formed different microstructures. Recall that microstructure is linked to how the alloying and atoms are arranged within the steel. Although we cannot realistically see the atoms, we can examine steel microstructure under a conventional microscope and use this to explain the effects of heat treating.

When the bolts were air cooled during normalizing, they formed the microstructures illustrated in **Figure 2.4**. The carbon separated into darker colonies called 'pearlite'. The carbon makes the pearlite notably stronger than the white 'ferrite' phase. Ferrite is essentially carbon-free which is why it is the weakest of all the phases. The more carbon in the steel, the more pearlite that forms and the stronger the steel. The higher pearlite content is what made the normalized Grade 5 bolt (0.4% C) stronger than the Grade 2 bolt (0.15% C).

The Grade 5 and 9 bolts had the same carbon additions and therefore, had similar pearlite contents. It was the molybdenum, chromium and nickel alloying that had further strengthened the Grade 9 bolt. Although their effects cannot be seen under a microscope, such alloying strengthens all microstructures including ferrite and pearlite. Yet recall that the Grade 9 bolt strength could be doubled by quenching and tempering – illustrating that the benefits of these expensive additives are maximized in the quenched and tempered condition rather than in the normalized state.

The dramatic strength increases experienced by the Grade 5 and 9 bolts from quenching and tempering were due to the formation of a

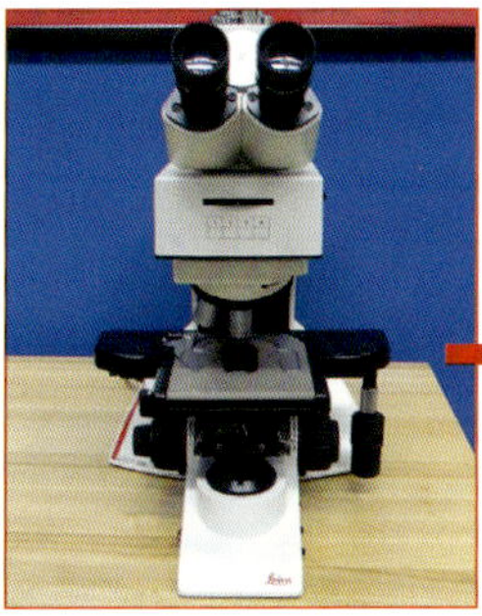

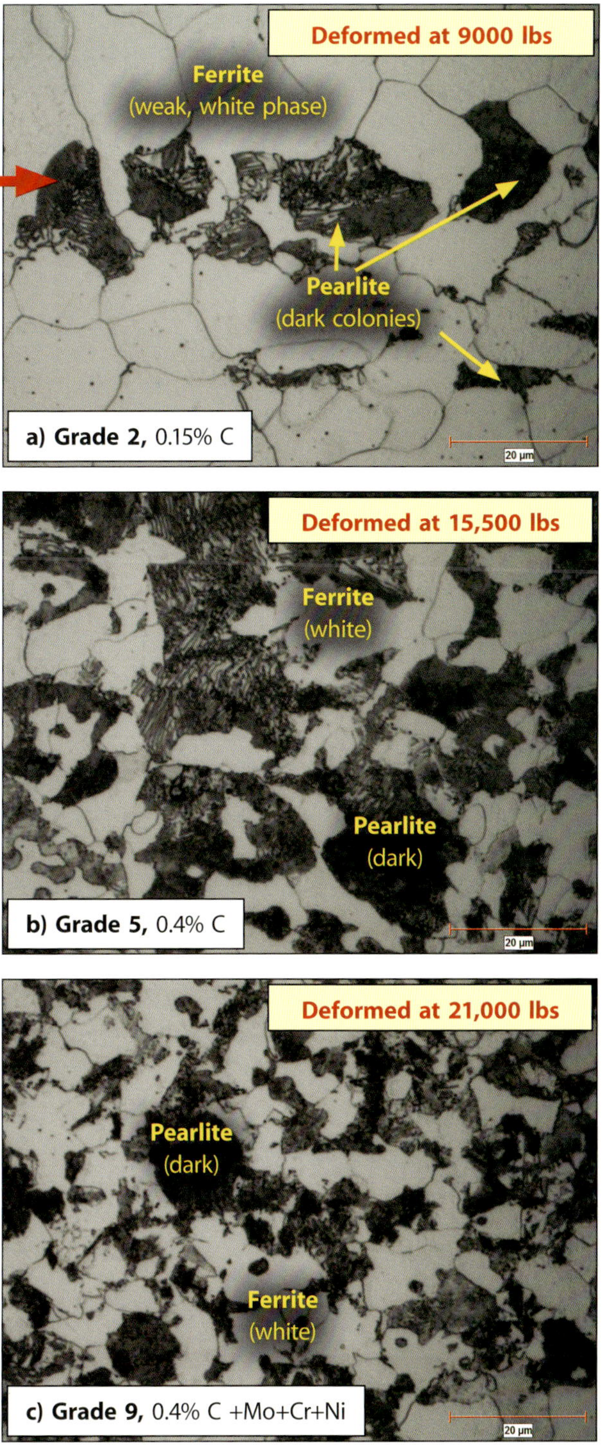

Figure 2.4: The Grade 2, 5 and 9 bolts after normalizing as seen through a microscope. Air cooling during normalizing had formed microstructures of ferrite (white) and pearlite (dark). The dark pearlite contains most of the carbon and is stronger than the ferrite. Images taken at 1000x magnification.

very different microstructure called 'tempered martensite'. **Figure 2.5** displays how quenching and tempering formed small, black dot carbides evenly dispersed throughout their microstructures. Carbides are a strong, carbon-rich phase. These dispersed carbides were the principal reason for the bolts' high strengths. Varying the tempering temperature affected the size and distribution of the carbides, altering their final properties.

Not only is tempered martensite stronger than ferrite and pearlite, it is harder. This is why steels with enough carbon and/or other alloying to form martensite are referred to as 'hardenable'. The reason the Grade 2 bolt did not have as significant a strength increase from quenching was that it did not have enough carbon to form martensite. It was not hardenable. Instead, it formed an 'upper bainite' structure that included the weak ferrite phase (**Figure 2.6**). Without the dispersed carbides of tempered martensite, it will never be as strong as hardenable steels quenched into martensite.

2.4 Effects of Alloying During Welding

Looking at the strengths and costs, one might wonder why we ever use low carbon steel like that of the Grade 2 bolt. However, low carbon steels are the most commonly produced steels in the world. We make bridges, buildings, freighters and refineries out of them. Their low carbon makes them easy to weld because they do NOT form martensite. Remember, quenching hardenable steels forms martensite which, if left untempered, is brittle. Welding steels with ample carbon and alloying can get complicated and requires additional steps to prevent the steel from becoming embrittled or cracked.

For example, when welds were made on the three bolts without any precautions, the Grade 5 and 9 bolts became brittle. Whereas the Grade 2 bolt remained ductile and bent when hammered, the Grade 5 and 9 bolts snapped apart under the same impacts (**Figure 2.7**). The heat from welding had changed the microstructure around the welds

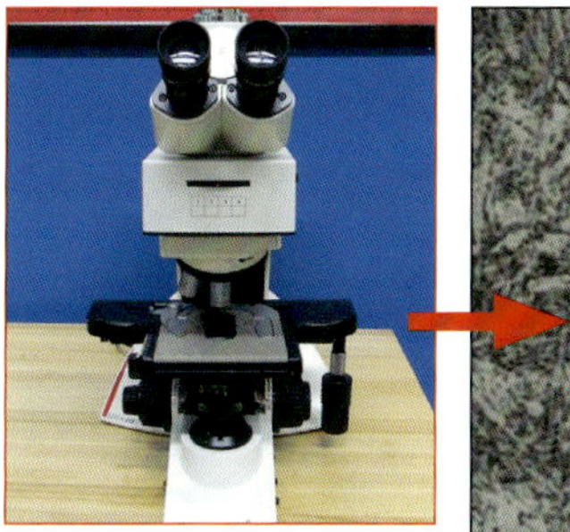

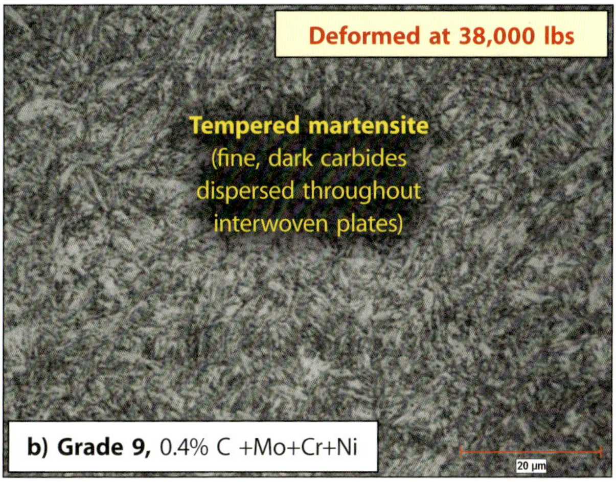

Figure 2.5: The Grade 5 and 9 bolts had formed tempered martensite microstructures from quenching and tempering. Fine carbides (dark dots) were dispersed throughout the martensite structure, providing high strength. These steels are 'hardenable' and able to form martensite because of their carbon and alloying contents. Images taken at 1000x magnification.

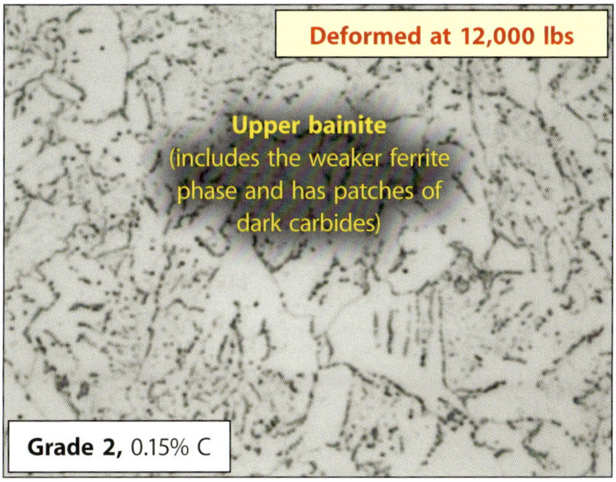

Figure 2.6: Micrograph illustrating how the Grade 2 bolt had not formed martensite during quenching. It had formed 'upper bainite' instead. Without the evenly dispersed carbides associated with tempered martensite, non-hardenable steels will never be as strong.

and the hardenable steels had become brittle (**Figure 2.8**). It is not that hardenable steels are not weldable, it is just that additional steps are needed to ensure their welds are durable and safe.

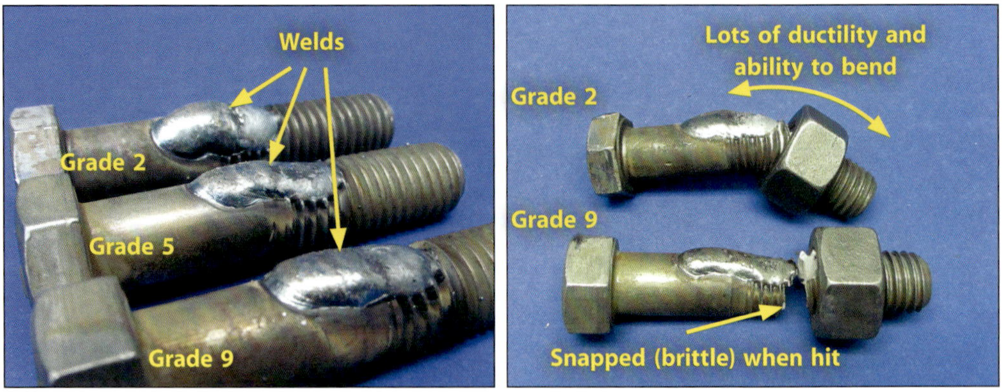

Figure 2.7: Each of the Grade 2, 5 and 9 bolts were welded and then hammered. The Grade 2 bolt retained its ductility and bent without cracking. However, the weld regions of the Grade 5 and 9 bolts had become brittle and snapped when hit.

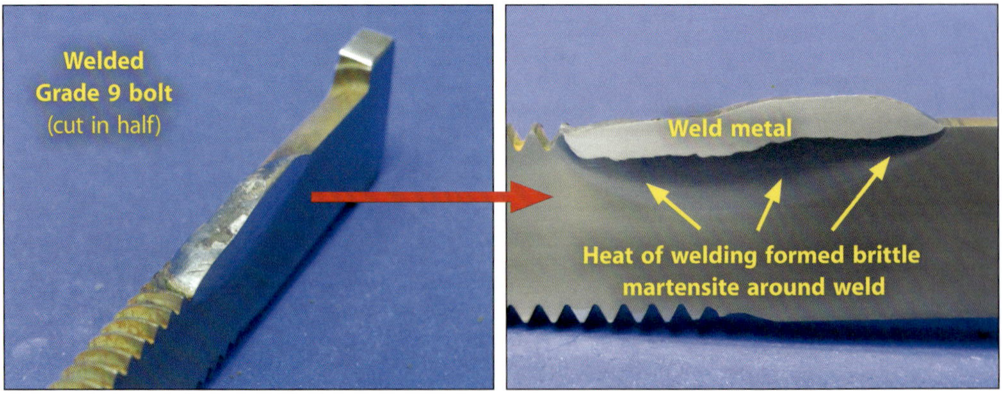

Figure 2.8: The steel surrounding the weld on the Grade 9 bolt had become hot enough to change its microstructure. Immediately after welding, the colder mass of the rest of the bolt quenched the weld region. Since the Grade 9 bolt was hardenable, this heating and quenching formed brittle martensite around the weld. It was this brittle martensite that had made it dangerously easy to break. Hardenable steels can be welded but they require extra precautions and take longer to weld compared to low carbon steels.

2.5 Summary of the Key Concepts

Understanding the effects of alloying and the microstructures that can be formed by heating provides the foundation to understanding steel behavior. The key principles can be summarized as:

1. Steel properties are based upon (a) its alloying and (b) the microstructure established during heat treating, welding and other processing.

2. Carbon is the primary alloying element of steel. It strengthens steel. If enough carbon is present, steel becomes 'hardenable' and has the ability to form martensite during quenching.

3. Normalizing by slow air cooling from red-hot temperatures forms microstructures of ferrite and pearlite. This provides moderate strength.

4. If hardenable steels are quenched from red-hot temperatures and then tempered, they can achieve high strengths by the formation of tempered martensite. Final properties can be adjusted by the tempering temperature.

5. Low carbon, non-hardenable steels are easy to weld because they do NOT form martensite. Higher carbon, hardenable steels are more complicated to weld because their welds can form brittle martensite, crack and be unsafe if precautions are not taken.

Although each of these concepts require more discussion, this overview provides the basis for selecting, designing, manufacturing and repairing steel components. Throughout the rest of this book, we will build on these concepts to provide a working understanding of steel.

Steel parts must be strong enough to endure the loading experienced

3.0 STRENGTH, DURABILITY AND HARDNESS

3.1 Strength

No matter what is being built – bridges, buildings, trains, cars, etc. – all components must be strong enough to bear the loading they will see throughout their service. This is why strength is the cornerstone property of steel. Although there may be several other properties relevant to each application, the first and foremost consideration is whether the steel will be strong enough.

The primary test to determine steel strength is 'tensile testing' (**Figure 3.1**). Steel is machined into a tensile specimen and then slowly pulled apart. The amount of load needed to stretch and cause failure is carefully measured. The stronger the steel, the more load needed to deform and tear it apart.

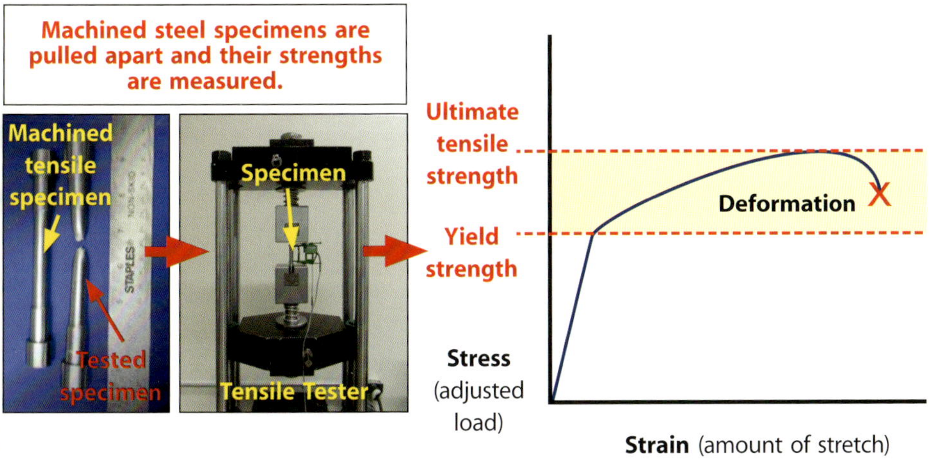

Figure 3.1: Tensile testing slowly pulls apart machined specimens while measuring their behavior. This generates a 'stress-strain curve' that provides the key strength values for that particular steel. The stronger the steel, the more load that is required to deform and break it.

In Chapter 2, we had simplified strength into how much weight a bolt could hold before it began to deform. Clearly the size of the bolt, or of any part, would affect how much weight would be needed to deform it. Bigger bolts made of the same strength steel would be able to hold more weight. Therefore, to reference solely the steel strength, regardless of part size, the units of 'stress' are used. Stress factors out the part geometry by dividing the load by the area over which the part is loaded.

Perhaps it is easier to illustrate what stress is than to describe it. **Figure 3.2** shows how the 9000 lbs load that caused permanent deformation of the normalized Grade 2 bolt was converted into the stress of 40.9 ksi (ksi - thousand pounds per square inch). If the same steel was made into bolts twice as large or half the size, its 40.9 ksi stress threshold makes it easy to calculate the weight that would cause each to deform – 18,000 lbs and 4500 lbs respectively (40.9 ksi multiplied by the loaded area). Using stress simplifies design and the evaluation of steel strength.

Comprehending the math is not needed to understand the role of stress and its importance will become clearer over time. For now, just remember that the higher the stress a steel can bear, the stronger the steel.

Once the data from tensile testing is adjusted to stress, it provides three important properties:

Yield Strength (YS) – The stress needed to cause permanent deformation.

Ultimate Tensile Strength (UTS) – The stress that breaks the steel apart.

Elongation – How much the steel stretched before breaking. Represents ductility.

Each of these properties play important roles in the design, manufacture and use of steel. Yield strength is the stress a particular steel can sustain before it begins to deform and irreversibly change shape. If loading on a part exceeds its yield strength, the part will distort and may cease to function properly, sometimes making it dangerous. Phrased differently, yield strength is the maximum one-time stress a part can bear before it begins to sustain damage. The steel may flex and bend a little under load but so long as the part operates below this stress threshold, it will spring back to its original shape when unloaded.

During the design phase, an engineer must anticipate the maximum stress a part will experience and ensure this is safely below the yield strength of the steel selected. Steel used during manufacturing must meet its design strength requirements. Through proper design, manufacturing and provided the part is used properly, the part should never reach or exceed its yield strength.

For example, all car parts are designed to operate well below their yield strengths. Throughout years of regular service, the maximum loading applied to gears, shafts, bearings, pistons

Figure 3.2: Steel strength is discussed in terms of 'stress'. This allows for the discussion of strength regardless of the part size. Stress is calculated by dividing the applied load/ weight by the area that had been loaded. For example, the normalized Grade 2 bolt began to permanently stretch/deform after 9000 lbs (pounds) of load. Dividing the load by the bolt cross-section size gives a stress of 40.9 ksi (kilo pounds per square inch).

and other engine components must remain below their yield strengths. If any of these components suffer a single loading event above their yield strengths (such as during a crash), they will become bent, buckled, distorted, twisted and/or stretched which will cause the drivetrain to seize. Even suspension components such as springs, that are designed to flex regularly, operate below their yield strengths and spring back to their original shapes once unloaded.

Ultimate tensile strength (UTS) is the amount of stress that will rip a particular steel apart. It is the measure of the steel's absolute strength. Since steel should never be loaded above its yield strength, something must have gone quite wrong for it to reach its UTS. However, in the event of a car crash that applies loading exceeding the yield stress and causes the car to buckle, higher UTS values will help increase the safety for the passengers inside.

Elongation is proof of steel ductility and is the percentage of how much the steel stretched during tensile testing. If manufacturing requires shaping by forming or bending, the steel must have enough ductility to form to the intended shape. Elongation is also an important measurement to prove that, combined with proper steel strength, the steel had been processed properly. No one wants to be surprised that a part had been unintentionally brittle.

Appendix A includes additional discussion regarding how steel behaves under load as described by the stress-strain curve. Yet it would be best to finish the main content of this book before reading the optional, more technical appendices.

3.2 Durability and Fatigue Limit

The yield strength represents the highest one-time load steel can sustain before deforming and distorting. Yet most steel parts need to survive more than one loading iteration. Engine components, buildings, bridges, trains and pumps all must endure decades of service and perhaps billions of loading cycles without breaking. If the repetitive loading were too close to the yield strength, cracks could form and grow to cause failure.

Ultimately, the repetitive loading a part can indefinitely sustain is well below the yield strength.

Steel has a stress threshold called its 'fatigue limit'. Below this stress level, it does not matter how many loading cycles a part experiences, it will not fail due to repetitive loading. Most steel parts expected to sustain millions of loading cycles are designed to operate safely below this stress level. The fatigue limit of steel is about half its ultimate tensile strength (UTS, **Figure 3.3**). This is another reason to ensure that steel has the intended UTS; it will ensure the steel has the intended fatigue limit and durability to survive cyclic loading conditions.

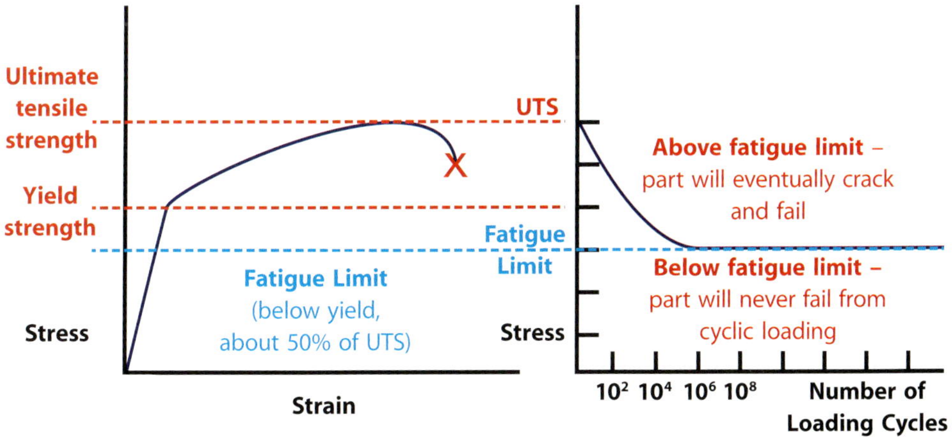

Figure 3.3: (Left) Stress-strain curve from a single loading iteration during tensile testing. (Right) Diagram illustrating how, even after millions or billions of loading cycles, there is a stress threshold below which cracking will not occur. This 'fatigue limit' is typically about half the UTS. Steel can survive an infinite number of loading cycles beneath this stress threshold. Steel parts that experience cyclic loading, like engine parts or bridges, are designed to never exceed their fatigue limits.

3.3 Hardness

Steel can be hard. Not hard to learn but physically hard. Hardness is the measure of how resistant a material is to indentation and scratching. Hardness also proves useful in another way – it is correlated to a steel's ultimate tensile strength (UTS). The harder the steel, the stronger it will be. If parts or features are too small to tensile test, hardness testing is a good second option to approximate whether the steel has the intended strength. Plus, hardness testing is comparatively fast and cheap to perform – making it a useful quality check. Few facilities have the ability to complete tensile testing yet many can perform hardness checks.

Most types of hardness testing machines work on the same principle. They push a hard indenter into the steel under a precise load and measure either (a) how deep the indenter impresses or (b) the indent size. The deeper or larger the impression, the softer the steel.

Rockwell hardness testing is the most common testing method used in manufacturing. **Figure 3.4** illustrates how a Rockwell hardness tester works. It reports steel hardness values using Rockwell scales B and C – abbreviated as HRB and HRC respectively. The HRB scale is used for steels in softer conditions such as after normalizing. When steel is harder than the HRB scale, usually from quenching and tempering, its hardness is measured with the HRC scale. Brinell (HB) is another hardness scale that is routinely used in the oil, gas and energy industries.

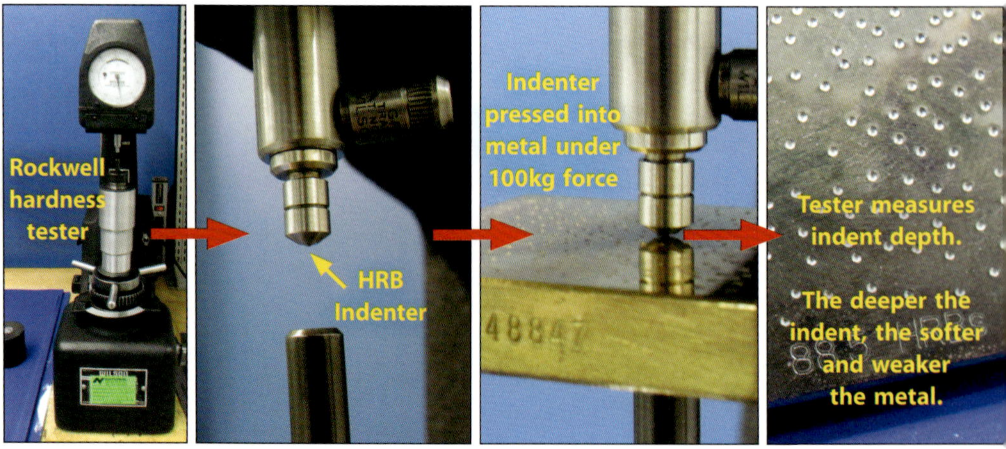

Figure 3.4: Photographs displaying a Rockwell hardness tester. A precisely shaped indenter is pushed into the metal under a specific load. The tester measures how deep the indenter impresses into the metal, reporting the corresponding hardness value. The deeper the indent, the softer the metal.

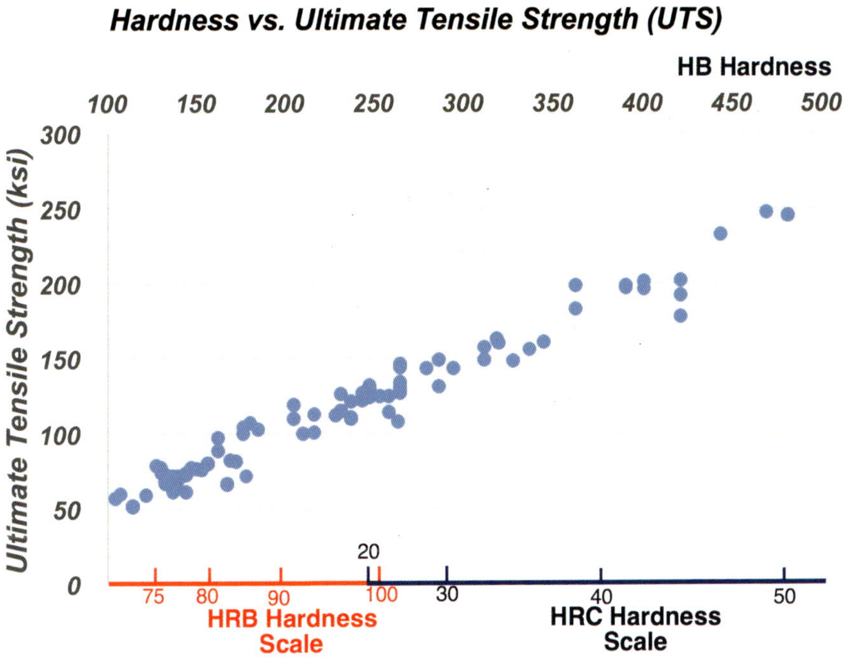

Figure 3.5: A steel's hardness is correlated to its UTS. The stronger the steel, the harder it will be. And vice-versa.

Figure 3.5 compares steel strength with hardness using the Rockwell and Brinell scales. Note the good correlation between hardness and strength. If a steel part cannot be tensile tested, hardness testing is the best alternative to ensure that it had been processed properly and is within the expected strength range.

3.4 Steel Compared to Other Metals

The reason why steel is so widely used is because of its combination of strength and affordability. It has the best cost-to-strength ratio of all conventional metals. To demonstrate this, a variety of bolts made of different metals including copper, brass, bronze, aluminum, titanium, stainless steel and nickel were purchased (**Figure 3.6**). Some bolts were of different sizes (some were metric) so their prices were not perfect

Figure 3.6: Numerous bolts made of various metals were purchased to illustrate the differences in costs versus strengths.

measures of the material costs. Still, once tensile tested, these bolts provided a clear demonstration as to why steel is the most commonly used metal in the world.

Table 3.1 ranks the various bolts from least to most expensive and lists their measured strengths. The cheapest bolt was the Grade 2 low carbon steel bolt. Although not overly strong for steel, this cheap bolt was stronger than the aluminum, copper, bronze and brass bolts that cost over five times as much. The second and third cheapest were the Grade 5 and 9 steel bolts – these were the strongest bolts tested.

Many of these other metals have important traits such as corrosion resistance, light weight or conductivity that justify their high price for particular applications. Yet when based solely on strength, no other metal comes close to being as cost-effective as steel.

Table 3.1: Comparing Bolt Cost and Strength of Various Metals

Metal Type	Major Alloying	Cost (USD)	Yield Strength (ksi)	UTS (ksi)
Steel - Grade 2 (low carbon)	Iron base 0.15% carbon	$0.45	70	80
Steel - Grade 5 (medium carbon)	Iron base 0.40% carbon	$0.60	130	145
Steel - Grade 9 (low alloy)	Iron base 0.40% carbon + Mo + Cr + Ni	$1.50	190	200
Stainless Steel (304)	Iron base 18% chromium 8% nickel	$2.38	125	150
Stainless Steel (316)	Iron base 17% chromium 11% nickel 2% molybdenum	$3.64	125	150
Aluminum (6061)	Aluminum base 1% magnesium 0.6% silicon	$3.69	40	45
Bronze (C65100)	Copper base 1.5% silicon	$4.40	65	70
Brass (C27000)	Copper base 35% zinc	$4.57	55	70
Copper	Commercially pure 99.9% copper	$5.61	30	35
Titanium (Ti6Al4V)	Titanium base 6% aluminum 4% vanadium	$9.66	120	130
Nickel Alloy (Monel 400)	Nickel base 30% copper	$33.12	95	110

3.5 Summary of Steel Strength, Durability and Hardness

1. Steel is cheaper yet often stronger than most other metals.

2. The two most critical strength properties are:

    a. Yield Strength – stress that begins to permanently deform the steel.

    b. Ultimate Tensile Strength (UTS) – stress needed to break the steel apart.

3. For steel parts expected to sustain repetitive loading, the stress levels need to remain below its 'fatigue limit'. The fatigue limit is usually about half the UTS.

4. Hardness is correlated to steel strength. Hardness testing is the best alternative to check the approximate strength when tensile testing is not possible.

Alloying is added and adjusted when steel is molten liquid. Once it solidifies, the steel chemistry is fixed. The alloying established will affect how the steel responds to processing and contribute to its final properties.

4.0 EFFECTS OF ALLOYING

Most steels comprise of over 98% iron. The careful addition of alloying to an iron base can provide an impressive range of useful properties. Alloying affects everything from strength to cost to what manufacturing processes can be used. Hence why an awareness of common additives and how they affect steel is useful for the designing, manufacturing and repairing of steel components.

4.1 Common Alloying

Table 4.1 summarizes the effects of the elements present within the two most prevalent families of steel: carbon steel and low alloy steel.

Chapter 2 discussed only the defining alloying of the Grade 2, 5 and 9 bolts. **Table 4.2** expands upon this and lists their detailed compositions. For carbon steels such as the Grade 2 and 5 bolts, five elements are reported. These five elements (C, Mn, Si, S, P) are stated for all steel families. For other steel types, any additional elements intentionally added or controlled are also reported – such as the molybdenum, chromium and nickel in the Grade 9 low alloy steel bolt.

Table 4.1: Summary of Common Alloying and Their Effects

Carbon Steel	
Five elements monitored within carbon steels. These five elements are in all steels.	
Carbon (C)	• Increases steel strength. Primary strengthening element. • Greatest impact upon strength of all alloying elements. • Above 0.30%, steels can form martensite if quenched (i.e. hardenable).
Manganese (Mn)	• Increases steel strength. • Similar but lesser effect as carbon. Has about 1/6 the effects of carbon.
Silicon (Si)	• Not overly relevant. Trivial increase in strength. • Added more for casting quality than for final strength properties.
Phosphorus (P) **Sulfur** (S)	• Impurities. Bad for the steel. • Efforts are made during steelmaking to minimize. • If each kept below 0.05%, not concerning.
Low Alloy Steel	
Low alloy steels contain the same five elements as carbon steel plus include at least one of the following:	
Molybdenum (Mo) **Chromium** (Cr) **Nickel** (Ni)	• Increases steel strength. • Maximum benefit achieved in quenched and tempered condition. • Increases size of part that can be quenched into martensite. • Reduces the amount of carbon needed to make steel hardenable.

Table 4.2: Detailed Compositions of the Grade 2, 5 and 9 Bolts

Bolt Grade	Steel Family	Composition (%)								
		C	Mn	Si	S	P	Mo	Cr	Ni	Fe*
		Strengthens steel. If enough present, makes it 'hardenable'.		Minimal strength effect.	Bad. Impurities. Not concerning if below 0.05%.		Strengthens steel.			Iron base.
2	Carbon Steel	0.15	0.60	0.04	0.011	0.015	-	-	-	balance
5	Carbon Steel	0.40	0.90	0.23	0.014	0.009	-	-	-	balance
9	Low Alloy	0.40	0.90	0.25	0.011	0.019	0.20	0.45	0.45	balance

*Fe (iron) is not usually reported – it is assumed that the remainder of the steel is iron.

Although iron makes up the vast majority of a steel's composition, this is assumed and the iron content is rarely declared.

All elements within steel play a role but some are more critical than others when first becoming familiar with steel.

4.2 Carbon

Carbon is the primary alloying element of steel. The difference of 0.2% carbon may seem trivial yet it is enough to dramatically change its properties, alter the appropriate manufacturing methods and change what applications the steel is suitable for. That difference could be enough to cause embrittlement during welding or result in considerably lower strength.

Carbon affects the steel in two ways:

1. Carbon increases the strength potential of steel. The higher the carbon, the greater its strength potential.
2. Above 0.30% carbon, steel becomes 'hardenable' and can form martensite when quenched.

Carbon additions provide increased strength at minimal financial cost. It was the higher carbon content that gave the Grade 5 bolt (0.4% C) greater strength than the Grade 2 bolt (0.15% C) in the normalized conditions. Plus, the higher carbon content of the Grade 5 bolt made it hardenable – providing a substantial strength increase by forming martensite when quenched and tempered. **Table 4.3** summarizes how the differences in carbon affected the microstructures formed and the strengths achieved by both (a) normalizing and (b) quenching and tempering heat treatments.

For parts that require minimal welding and limited forming during manufacture (shafts, pistons, gears, bearings, etc.), it may be ideal to use steel with ample amounts of carbon. Yet abundant carbon comes with a less obvious cost – it can complicate manufacturing and add to the production expense. For example, welding steel with over 0.30% carbon requires additional steps that add time and cost. This is why large constructions that require lots of welding (buildings, bridges, refineries, freighters, etc.) are made of low carbon steels with less than 0.25% carbon, safely below the hardenable threshold.

Table 4.3: Effects of Carbon on Strength and Hardenability

Bolt Grade	Carbon (%)	Normalized (air cooled)			Quenched Then Tempered at 900°F		
		YS (ksi)	UTS (ksi)	Microstructure	YS (ksi)	UTS (ksi)	Microstructure
2	0.15	41	60	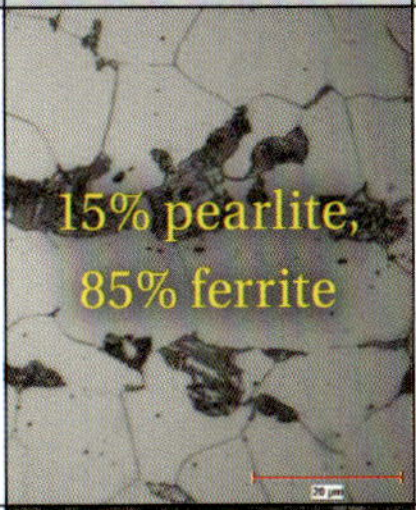	55	75	
5	0.40	70	105		140	155	

YS – Yield strength.
UTS – Ultimate tensile strength.

4.3 Molybdenum, Chromium and Nickel

The addition of molybdenum, chromium and/or nickel can further increase the strength of steel. More specifically, they increase the strengths of all microstructural phases including ferrite, pearlite and martensite. Recall that the Grade 5 and 9 bolts had similar carbon contents and both formed martensite when quenched and tempered. Yet the additional molybdenum, chromium and nickel gave the Grade 9 bolt higher strength over the Grade 5 bolt (**Table 4.4**).

These alloying elements also help promote the formation of martensite during quenching. Their additions reduce the amount of carbon needed to make a steel hardenable and make it possible to quench larger parts into martensite.

Table 4.4: Effects of Molybdenum, Chromium and Nickel on Strength and Hardenability

Bolt Grade	Primary Alloying				Quenched Then Tempered at 900°F		
	C	Mo	Cr	Ni	YS (ksi)	UTS (ksi)	Microstructure
5	0.40	-	-	-	140	155	
9	0.40	0.20	0.45	0.45	170	180	

Due to their relatively high costs, low alloy steels are most often used in the quenched and tempered condition where their maximum benefits are achieved.

4.4 Other Steel Families and Alloying

There are other steel families that have different alloying and microstructural strategies. **Appendix B** includes examples of common steel grades used in industry today, expanding the discussion to include additional steel families. It is best to read the main text of this book before considering whether to read the optional appendices.

4.5 Analyzing the Steel Composition

Considering how important alloying is to steel, it is critical that steels used during manufacture have the correct compositions. Steel is purchased to specific alloying limits and should come with a 'material certificate' stating its exact composition (further discussed in Chapter 6). If such documents are unavailable or verification is needed, there are several methods to analyze the composition of steel. The most common requires sending a small piece of steel cut from the part to a laboratory for optical emission spectroscopy (OES) testing. This burns the steel surface to determine its alloying content (**Figure 4.1a,b**).

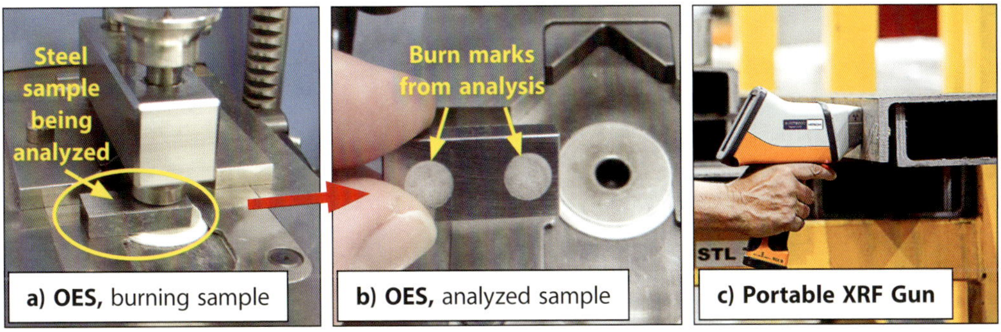

Figure 4.1: (a,b) OES uses an electric arc to burn the steel surface, evaluates the light emitted and uses elemental light signatures to determine the steel's composition. (c) A portable XRF gun uses X-rays to approximate the elements within steel. Although less accurate than OES, it can quickly analyze steel anywhere and without having to cut the part.

If the steel part cannot be cut or sent to a laboratory, another analysis method involves using a portable X-ray fluorescence (XRF) gun (**Figure 4.1c**). Although this method does not provide precise results, it can usually identify the type of steel.

4.6 Summary of Alloying

1. Alloying is adjusted when steel is molten liquid. Once it solidifies, its alloying is set.

2. Alloying affects how steel responds to processing and its final properties.

3. Carbon is the primary alloying element of steel:

    a. Carbon increases the strength potential of steel.

    b. Above 0.3% carbon content, steel becomes hardenable and high strength can be achieved by quenching and tempering.

4. Additional alloying such as molybdenum, chromium and nickel:

    a. Further increases the steel strength.

    b. Increases the part size that can be quenched into high strength martensite.

    c. Reduces the amount of carbon steel needs to be hardenable.

    d. Provides maximum benefit in the quenched and tempered condition.

5. Steel used during manufacturing must have the intended alloying. If steel type is unknown or the alloying needs verification, testing can be completed by:

    a. Sending a steel piece to a laboratory for OES analysis (accurate) or,

    b. Using a portable XRF gun (less accurate).

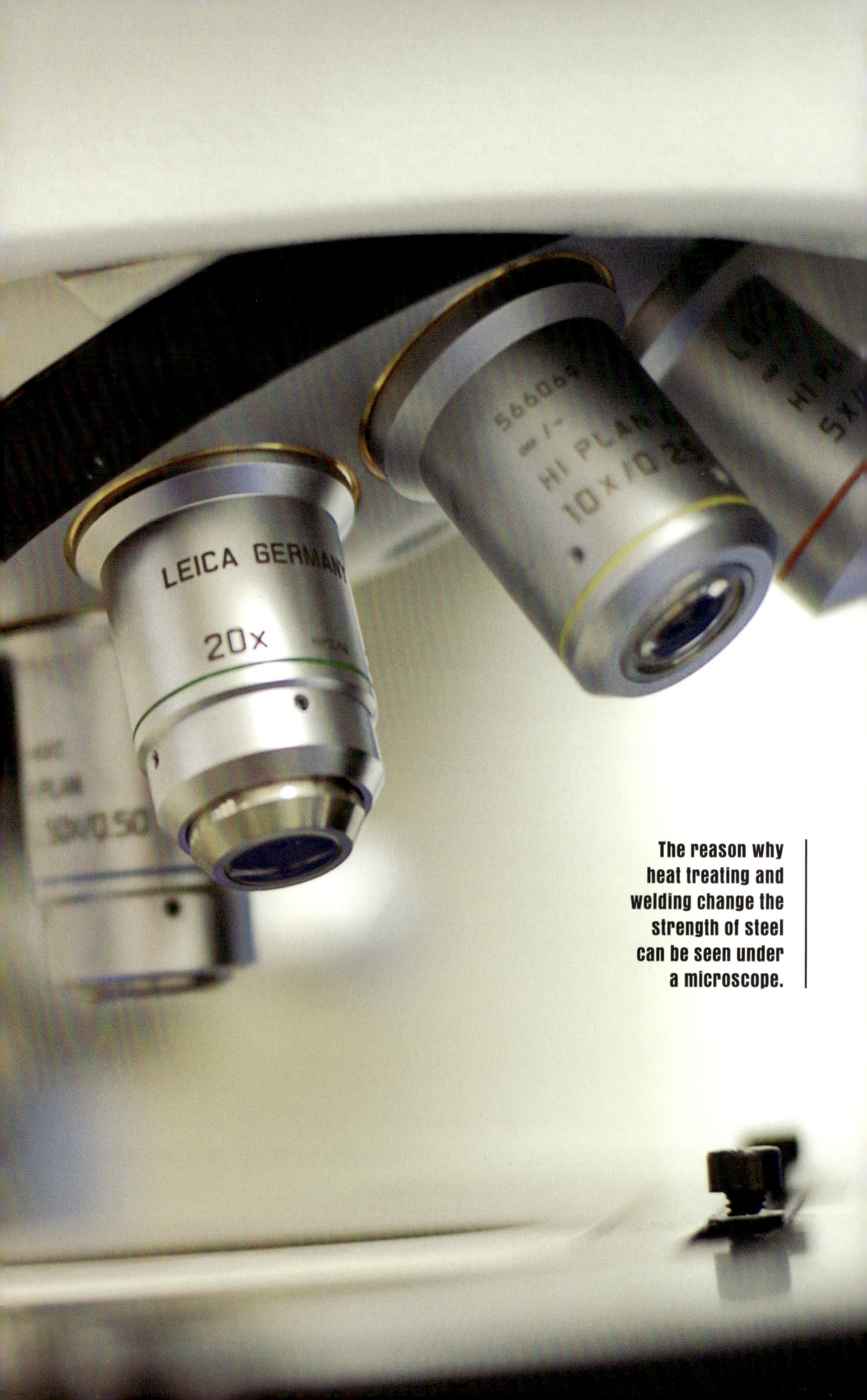

The reason why heat treating and welding change the strength of steel can be seen under a microscope.

5.0 MICROSTRUCTURE AND HEAT TREATMENT

Designing and manufacturing rely on either purchasing steel with the intended properties or being able to modify the steel to achieve those properties. In either case, the properties of steel will always be based upon (1) its composition and (2) the microstructure established through processing. This means that once the steel type and its alloying have been selected, all further efforts to ensure the steel has the intended properties are based upon manipulating its microstructure. As shown in Chapter 2, heat treatments and welding can dramatically alter the properties of steel by changing its microstructure. Understanding the traits of different microstructural phases and how they can be formed is useful for intelligent design and the efficient manufacture of robust components.

5.1 Observing the Microstructure

Microstructure represents how the atoms and alloying are organized within a particular steel. The ability to examine the microstructure of steel using conventional microscopes has allowed science to explain its behavior and help optimize the processes used by industry today.

Microstructure can be observed by polishing the surface of a small steel piece to a mirror finish, chemically etching it and then examining it under a microscope. Steel parts usually have to be cut down in size and are often mounted in plastic or epoxy before polishing. **Figure 5.1** shows a typical sequence used to examine steel microstructures.

Few companies handling steel have the laboratory facilities or interest to examine microstructures. Steel microstructures are usually only evaluated when developing a new process, as an occasional quality check or if a problem arises. Yet the only way to explain steel behavior and why processing affects its properties is to discuss the influence of microstructure.

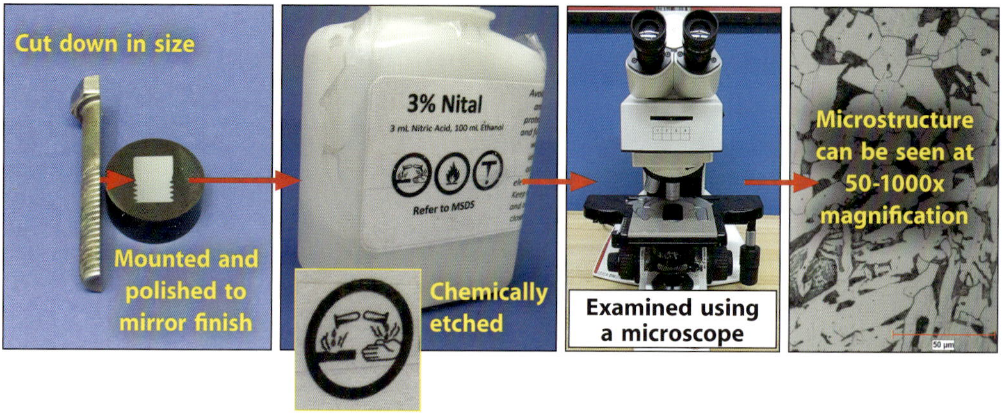

Figure 5.1: Before the microstructure of steel can be observed under a microscope, it needs to be prepared in a specific manner. The steel piece is cut and usually mounted into plastic or epoxy. It is polished to a mirror finish and then chemically etched. Unless noted otherwise, all images of steel microstructures within this book were chemically etched with dilute nitric acid called '3% nital'.

5.2 Resetting the Microstructure

A great deal of steel behavior depends upon its primary strengthening element – carbon. Following the migration of the carbon atoms during the different stages of heat treating and within the microstructure provides insight into what is happening.

The microstructure of steel is first established during processing at red-hot temperatures. At temperatures above 1340°F (727°C), the carbides of the former structure dissolve and the associated carbon becomes dispersed throughout the steel, effectively resetting the microstructure. This carbon dispersion is caused by the gaps between the iron atoms becoming large enough that the small carbon atoms can fit between them. Once steel begins to cool down, the gaps between the iron atoms shrink and the carbon atoms no longer comfortably fit. What microstructure is formed upon cooling and whether the carbon is squeezed out from the iron matrix depends upon how fast the steel is cooled and its alloying.

Upon cooling, there are four common microstructural phases or colonies that can form. These are listed in **Table 5.1**. Although these microstructures have already been quickly introduced in Chapter 2, their importance warrants further discussion.

Table 5.1: Common Microstructures

Microstructure	Established by
(1) Ferrite and (2) Pearlite	Slow cooling (normalizing, annealing, etc.).
(3) Martensite	Rapid quenching (in water or oil).
(4) Upper Bainite	Either (a) rapid quenching of non-hardenable steel or (b) insufficient quenching of hardenable steel.

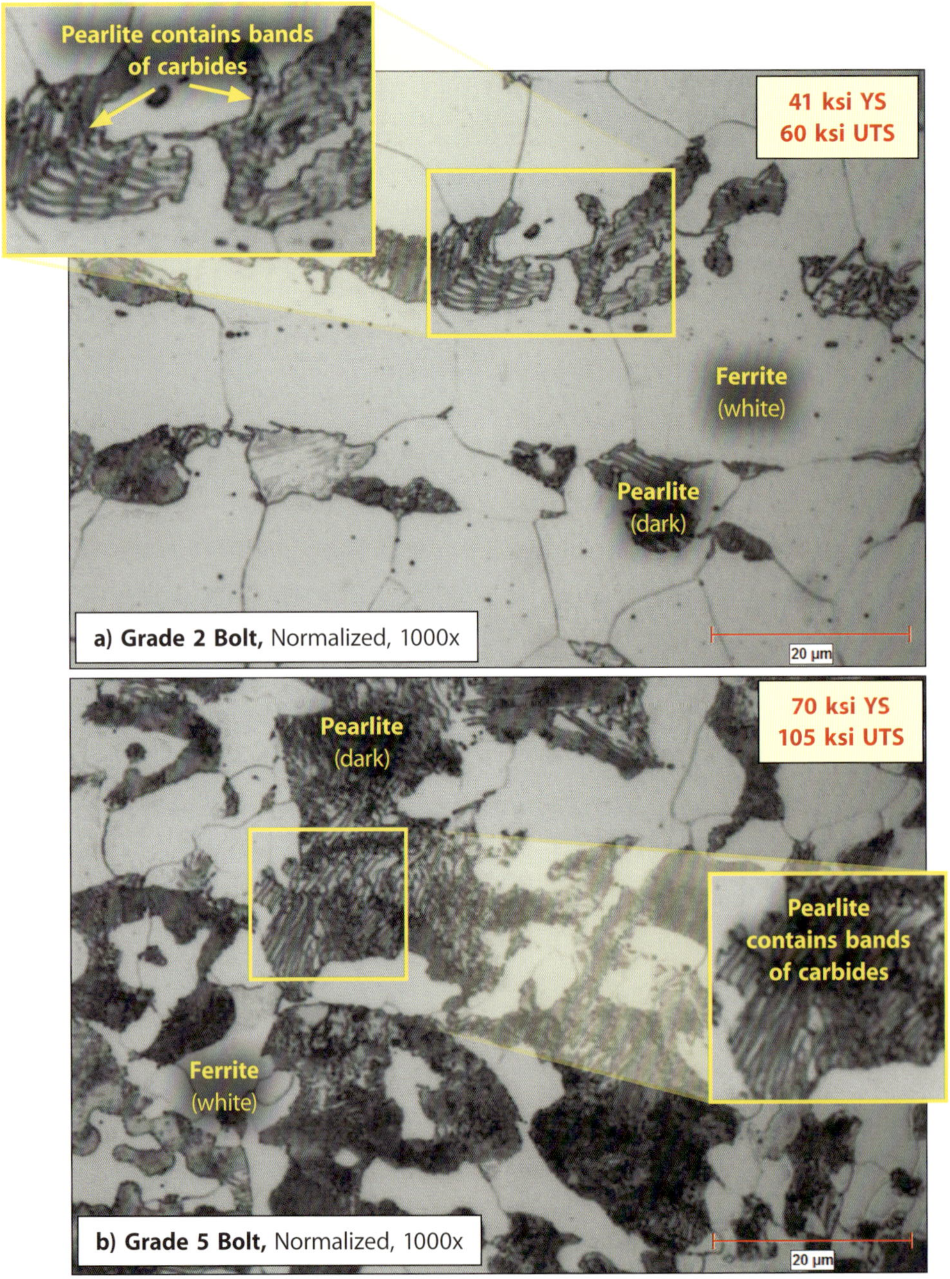

Figure 5.2: Micrographs displaying the ferrite (white) and pearlite (dark) microstructures of normalized, air cooled Grade 2 (0.15% C) and Grade 5 (0.4% C) bolts. Pearlite contains bands of carbides that provide higher strength than the carbon-free ferrite. Note the difference in strengths between the two bolts which was largely based upon their proportions of pearlite. The more carbon present, the more pearlite that will form and the stronger the steel.

5.3 Ferrite and Pearlite

Intermixed ferrite and pearlite is formed when steel cools from red-hot temperatures slowly enough that the carbon has time to escape from the iron matrix and separate into two different phases. **Figure 5.2** displays the ferrite and pearlite microstructures of the normalized Grade 2 and 5 bolts. The white ferrite phase has almost no carbon which makes it the weakest of all the phases. Most of the carbon in the steel separates into pearlite and forms bands of iron-carbon 'carbides'. These dense, fine carbide bands strengthen the pearlite. The more carbon in the steel, the more pearlite that forms and the stronger the steel.

As discussed before, 'normalizing' is the heat treatment used to reset the microstructure and establish a ferrite-pearlite structure. Normalizing is completed by simply heating steel to sufficiently hot temperatures for a period of time and then removing it from the furnace, allowing it to air cool (**Figure 5.3**). Most steels are normalized between 1600°F and 1700°F (870-925°C) with the precise temperature depending upon the steel composition.

Figure 5.3: Normalizing heat treatments are used to establish ferrite and pearlite microstructures. It includes heating steel parts to temperatures hot enough to reset their structures, holding for a period of time and then removing from the furnace to air cool. Different steel grades and part sizes require different temperatures and times to normalize.

Normalized steels generally exhibit moderate strength, good ductility and are easy to machine or shape. The exact properties established are influenced by the steel's composition which dictates the balance of ferrite and pearlite produced. The lower the carbon content, the more of the weak but ductile ferrite phase that will form. Steels with a predominantly ferrite structure can be readily shaped, welded and machined. Increasing the carbon and pearlite content increases the steel strength but reduces its ductility.

The reason why steels are normalized varies upon the application, its alloying and the preferred methods of manufacture. Normalizing (or another heat treatment) is often completed after shaping processes such as casting, hot forming or forging. Due to their exceptionally hot temperatures, these processes may form non-ideal microstructures and poor properties. Normalizing afterward ensures that a uniform microstructure is established that provides a balance of moderate strength and ductility.

There is another heat treatment that forms a ferrite and pearlite structure designed to optimize the ease with which steel can be machined and shaped. 'Annealing' intentionally softens and weakens steel by forming very coarse pearlite. Annealing begins the same way as normalizing – by heating steel to red-hot temperatures. Yet instead of air cooling, the steel is kept in the furnace and the entire furnace is slowly cooled. The slower the furnace cools the steel, the coarser the pearlite and the weaker the steel will become. This maximizes the steel's ductility and makes even highly alloyed steels easy to machine, shape and form. **Figure 5.4** compares the microstructures and strengths of normalized and annealed bolts. Structural parts are never used in this annealed, weakened condition. Once machined and shaped, load-bearing parts are then heat treated to their final condition.

For those interested in better understanding the connection between microstructure and atomic arrangement, **Appendix C** includes further reading on the matter.

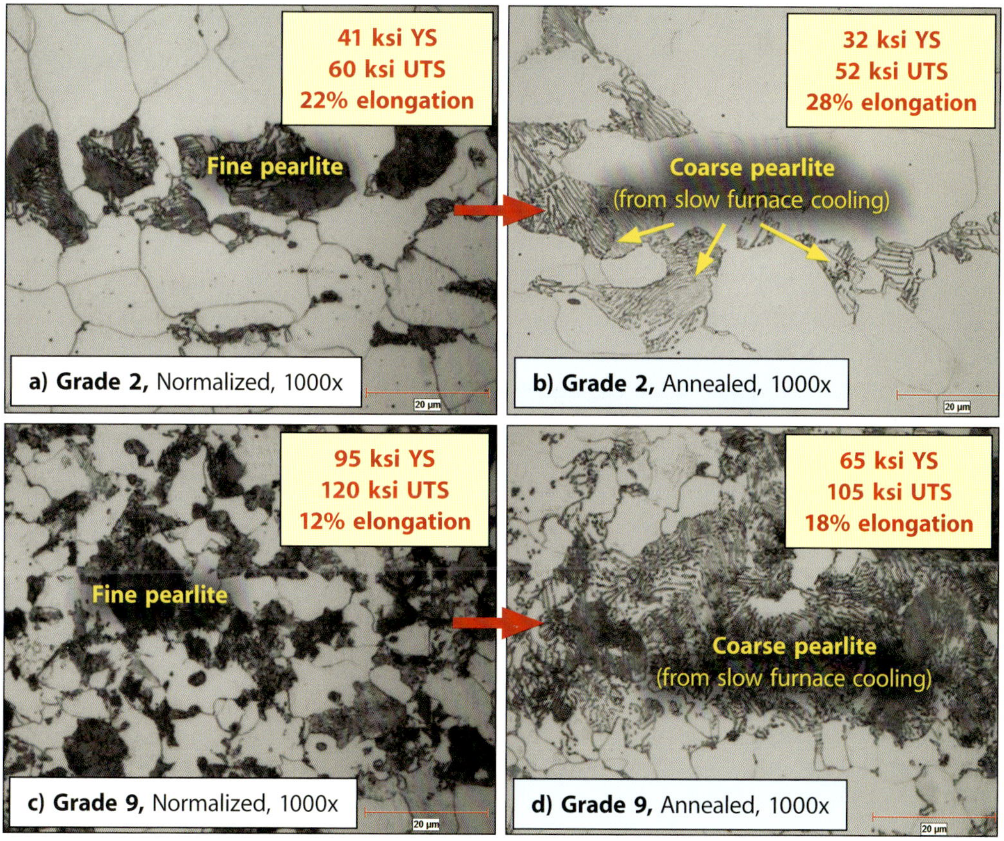

Figure 5.4: Micrographs comparing the microstructures of normalized and annealed Grade 2 and 9 bolts. The coarser pearlite formed during annealing provided less strength compared to the finer pearlite from normalizing. Low carbon steels are annealed to maximize their ductility, useful for forming them into complex shapes. High strength steels are annealed to make them easier to shape and machine yet are later heat treated again to reestablish their strengths.

5.4 Martensite

Steel becomes hardenable with 0.30% carbon content or, if additional alloying is included, at even lower carbon contents. Hardenable steels have the capacity to form martensite when quenched from sufficiently

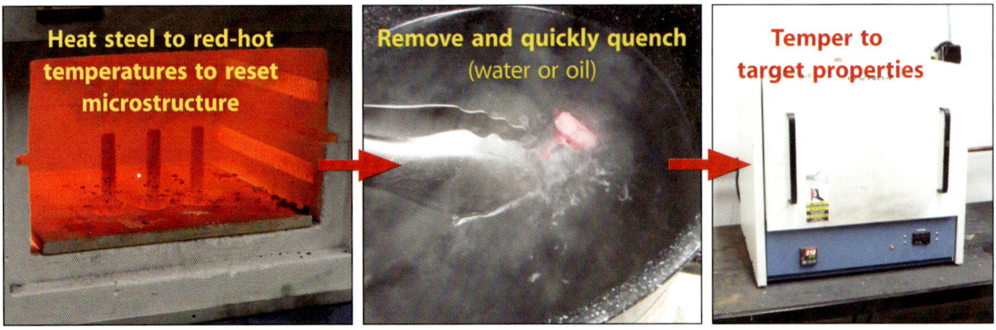

Figure 5.5: 'Quenching and tempering' includes heating steel parts to a sufficiently hot temperature, holding for a period of time, then removing and quickly submerging them in a quenchant such as water. The parts will then be tempered by heating to moderate temperatures. Different steel grades, part sizes and target properties require different temperatures and quenchants. Steel that has been quenched will sometimes be called 'hardened'.

hot temperatures. The formation of martensite provides the ability to achieve high strength.

Figure 5.5 briefly reviews the 'quenching and tempering' heat treatment used to establish a tempered martensite microstructure. The temperatures used to heat steel before quenching will vary based upon its composition yet is usually between 1550°F and 1600°F (845-870°C). Tempering temperatures used after quenching can range from 600°F to 1200°F (315-650°C) depending upon the target properties.

The rapid quenching from red-hot temperatures does not provide enough time for the carbon atoms, nestled between the iron atoms, to escape as everything quickly contracts. For hardenable steels, quenching and the sudden formation of martensite traps the carbon between the iron atoms. This as-quenched martensite is brittle because the trapped carbon prevents atomic movement – preferring to break rather than allow the atom shuffling needed to bend or deform. **Figure 5.6** displays an as-quenched Grade 5 bolt that snapped at a dangerously low stress because it could not accommodate any bending.

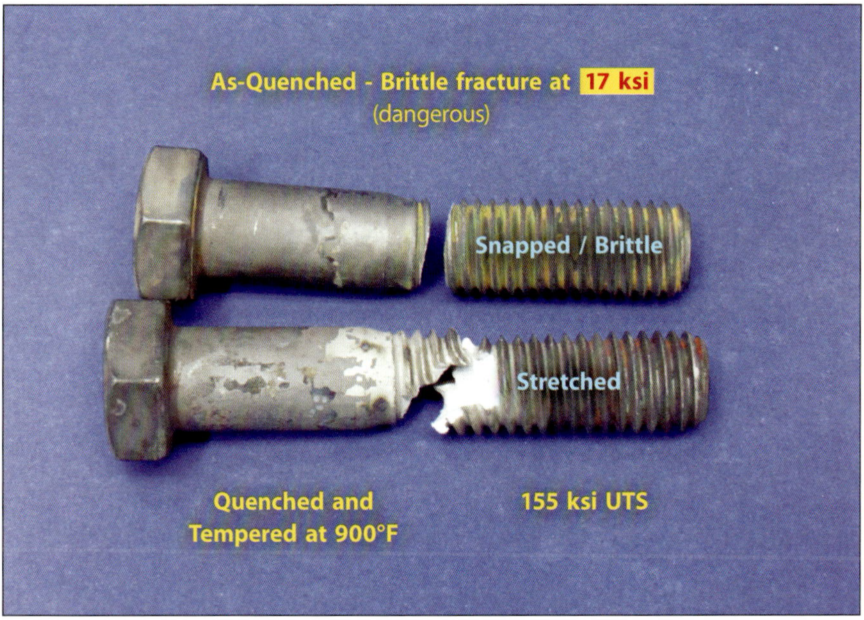

Figure 5.6: An as-quenched Grade 5 bolt (no temper) fractured at an alarmingly low stress level. When tensile tested at a slight angle, the as-quenched bolt snapped at a mere 17 ksi – this was about one-tenth the stress needed to break a tempered bolt. The as-quenched, untempered martensite was brittle and had no ability to bend or stretch. The microstructural change that occurs during tempering is needed to restore ductility and make the steel safe for use.

When quenched steel is tempered at moderate temperatures, the carbon quickly escapes from between the iron atoms to form small, round carbides. **Figure 5.7** displays the tempered martensitic structure of a Grade 5 bolt (0.4% C) after quenching and tempering at 1100°F (593°C). Tempered martensite is comprised of plates filled with evenly dispersed carbides. It is these fine carbides that strengthen the steel. Their size and density can be adjusted by the tempering temperature, which in turn modifies the steel strength and ductility (**Figures 5.8** and **5.9**). Hotter tempering temperatures reduce the steel strength yet improves its ductility – an important trait for some applications and manufacturing processes. Note that tempered martensite is rarely as ductile as normalized or annealed steels.

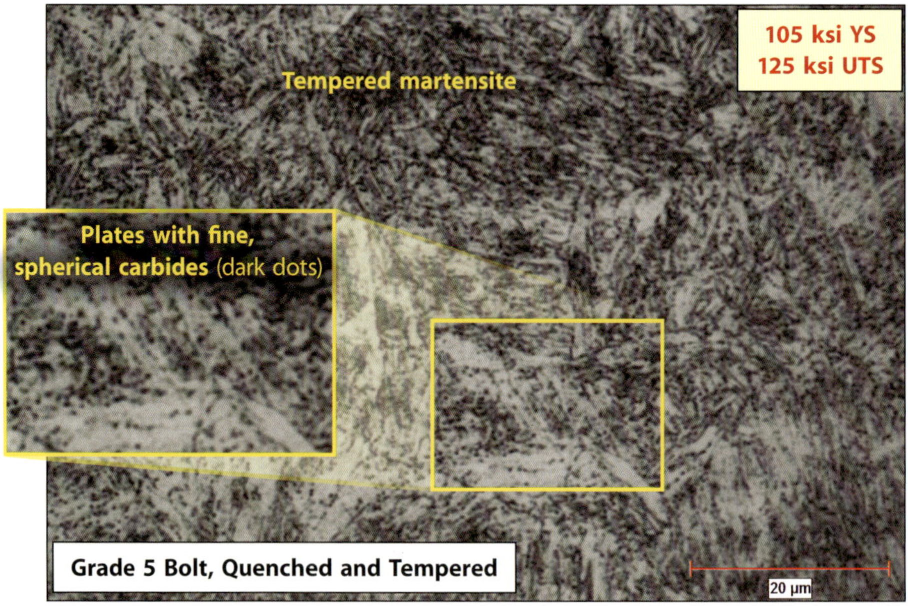

Figure 5.7: Tempered martensite comprises of needle-like plates with fine, dark carbides dispersed throughout. It is the dispersion of these fine carbides that provide most of its strengthening. Grade 5 bolt (0.4% C) quenched and tempered at 1100°F (593°C).

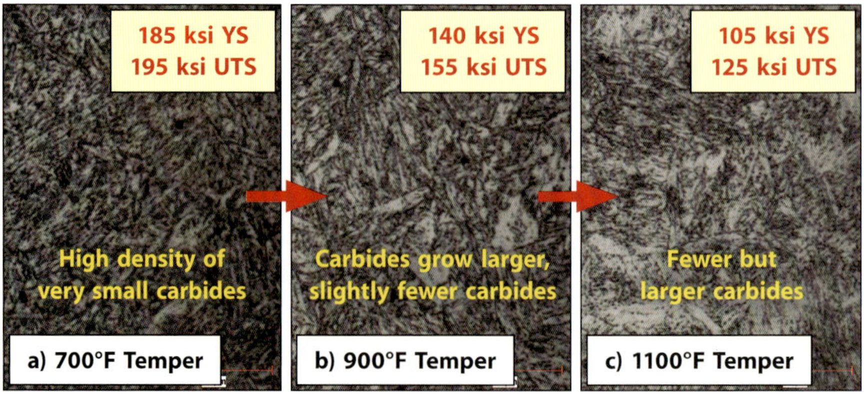

Figure 5.8: The tempering temperature affects the carbide size and distribution within tempered martensite. This in turn affects the steel properties. The higher the carbide density, based upon carbon content and tempering temperature, the more strengthening they provide. As higher tempering temperatures are used, the carbides become larger but less numerous and provide less strengthening. Grade 5 bolt (0.4% C) quenched and tempered.

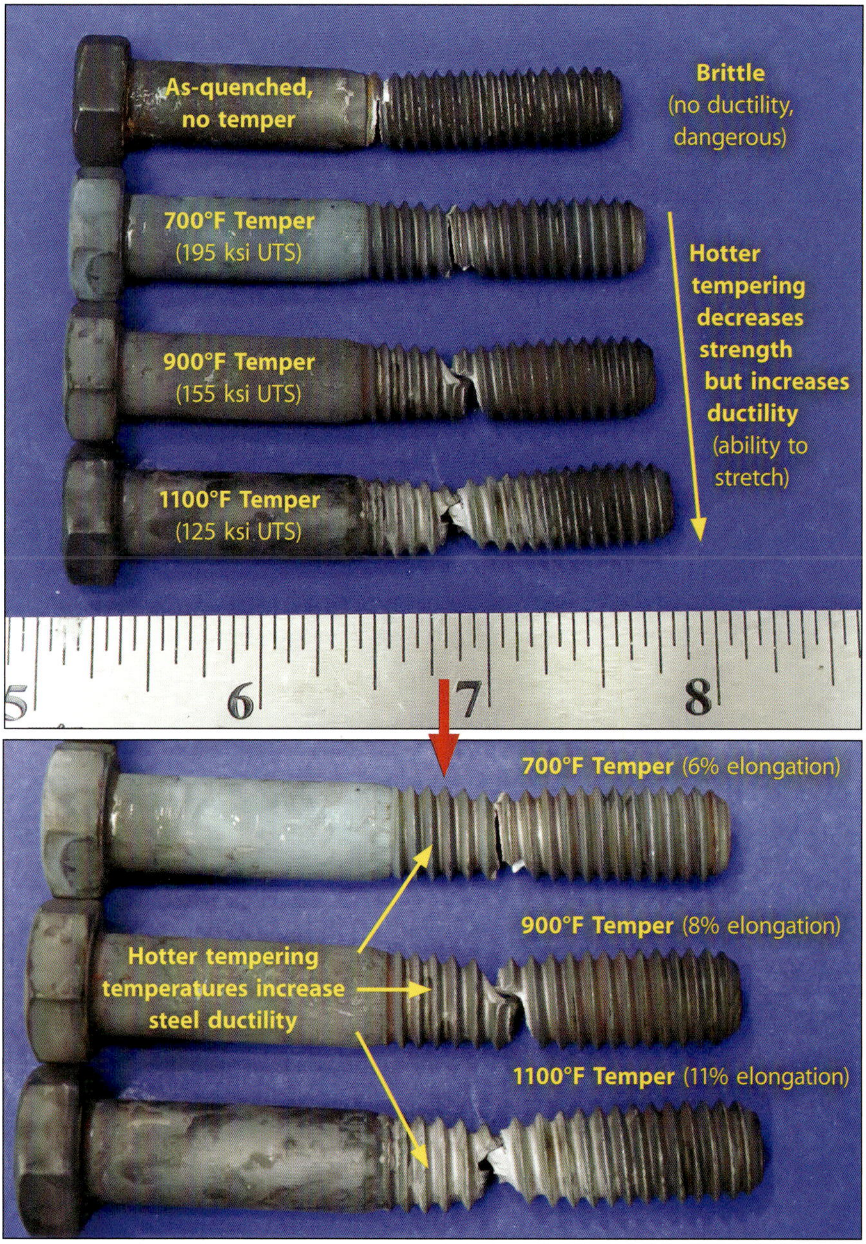

Figure 5.9: Tempering restores and adjusts the ductility of martensite. As demonstrated with Grade 5 bolts, increasing the tempering temperature increased the steel's ductility and how much it could stretch before breaking. Increasing ductility has benefits during manufacturing and provides added safety during overload incidents. This ductility comes at the expense of strength.

The reason why steel is quenched and tempered is obvious – it imparts high strength. This heat treatment and the formation of tempered martensite is what doubled the strengths of the Grade 5 and 9 bolts compared to their normalized conditions. Tempering can also be used to balance strength with other properties that might be important for a part, such as ductility and the ease with which it can be machined. Once the balance of ideal properties has been determined, ample literature exists regarding what properties are established by different heat treatment parameters and temperatures for each steel type.

5.5 Upper Bainite

There is an intermediate phase called 'upper bainite'. Upper bainite is rarely an intentional structure from heat treatment. It has a coarse structure that includes needles of ferrite (the weakest phase, **Figure 5.10**). Upper bainite forms when non-hardenable, low carbon steels are (a) quenched from red-hot temperatures or (b) if cooled from excessive, near-white-hot temperatures such as after forging. Hardenable steels can form upper bainite if they are not quenched fast enough to form martensite. With the inclusion of ferrite and without the evenly dispersed carbides to strengthen it, upper bainite does not have the same strength potential as tempered martensite.

Although upper bainite is stronger than mixed ferrite and pearlite, its final properties are overly sensitive to the cooling rate and cannot be trusted to provide consistent results. Hence why no common heat treatment intentionally forms upper bainite. Yet upper bainite is often formed during less controlled processes such as welding.

5.6 Hardenability

Hardenable steels have enough carbon and alloying to form martensite when quenched. 'Hardenability' refers to how easily a steel forms

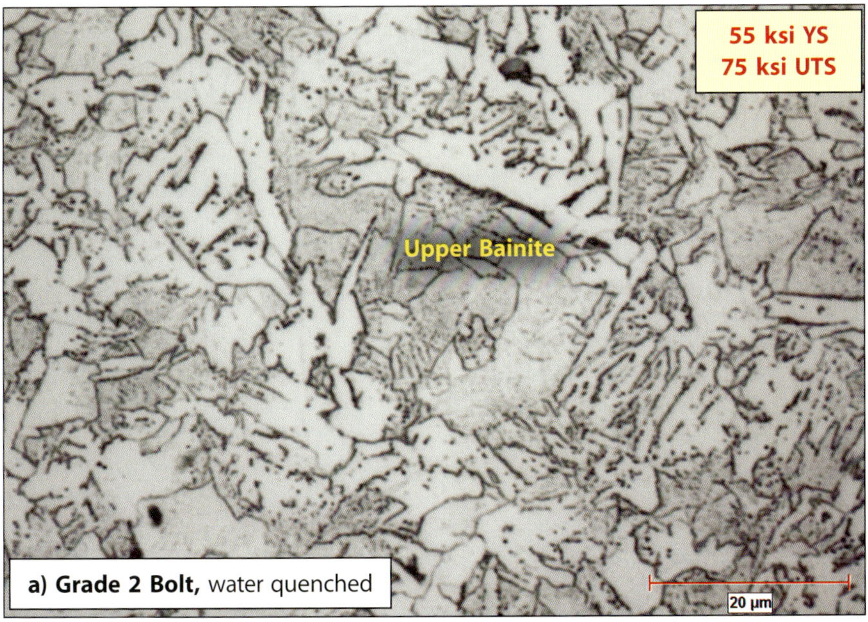

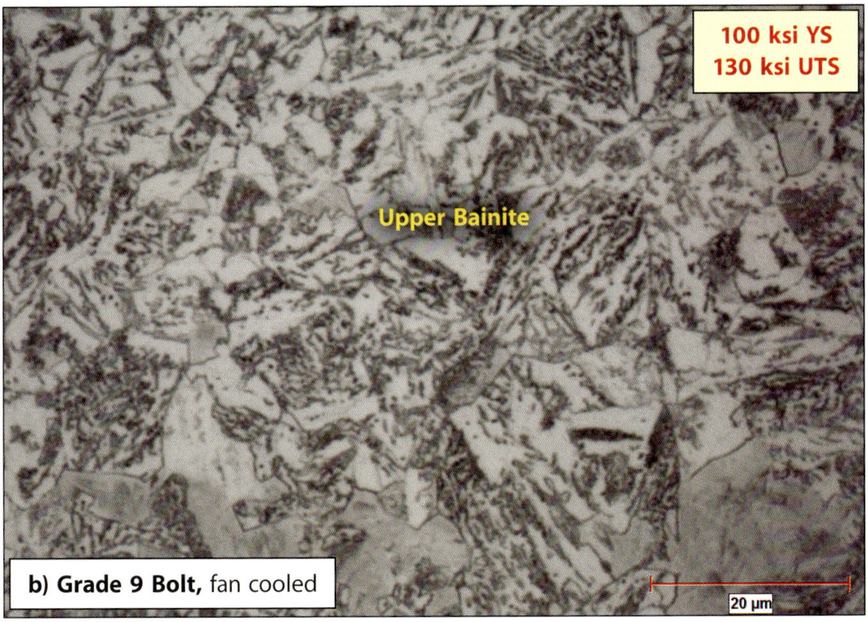

Figure 5.10: (a) Water quenching of a Grade 2 bolt (0.15% C) formed 'upper bainite This low carbon steel bolt did not have sufficient carbon and/or alloying to form martensite when quenched – so it formed upper bainite instead. (b) A Grade 9 bolt (0.4% C + Mo+Cr+Ni) had formed upper bainite from being fan cooled. Fan cooling had not cooled the bolt fast enough to form martensite.

martensite. The higher the hardenability, the slower a steel can be quenched and still form martensite. Since the cores of large parts will quench slower than that of small parts, increasing the hardenability enlarges the part size that can be 'through-hardened' entirely into martensite.

For example, **Figure 5.11** displays the cross-sections of bars ranging from ½-inch to 1½-inch in diameter that had been heated and water quenched. All three bars comprised of the same medium carbon steel with 0.45% carbon. Although this steel is hardenable, only the smallest diameter bar had been through-hardened into a fully martensitic structure. The surfaces of the two larger bars had formed martensite but their cores comprised of weaker phases (mixtures of ferrite, pearlite and bainite). Therefore, due to its low hardenability, only thin parts of this steel type can benefit from the strength increase associated with quenching and tempering.

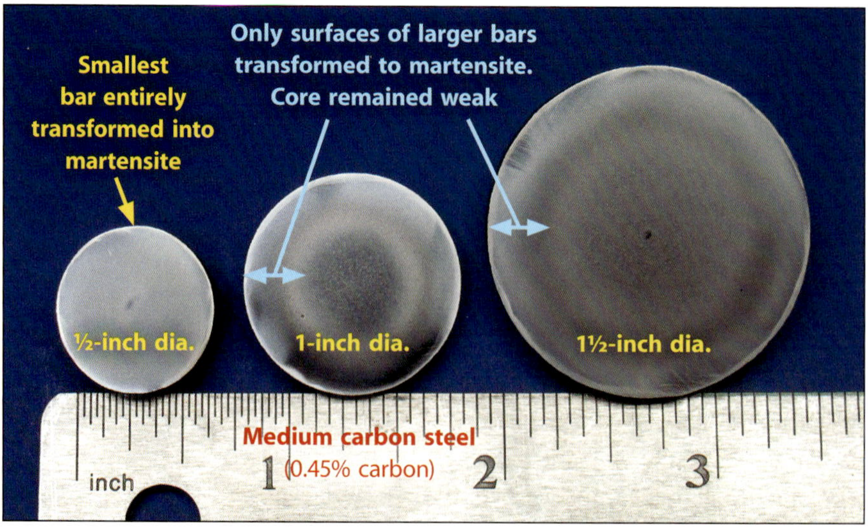

Figure 5.11: Etched cross-sections of bars made of 0.45% carbon steel, measuring ½, 1 and 1½-inches in diameter, after having been heated to 1570°F (855°C) and quenched. The smallest, ½-inch diameter bar had been through-hardened into martensite. Due to the steel's low hardenability, only the surfaces of the larger bars had quenched fast enough to form martensite, leaving the cores relatively weak.

The hardenability of steel can be improved by adding molybdenum, chromium, nickel and/or further increasing the carbon content. **Figure 5.12** shows how a 3-inch diameter bar of a low alloy steel had been through-hardened into martensite under the same conditions. This steel contained molybdenum and chromium which greatly increased its hardenability. High hardenability is needed to produce large, high strength parts. Therefore, the costs of molybdenum, chromium and/or nickel alloying may be justified for larger parts that require the strength of martensite.

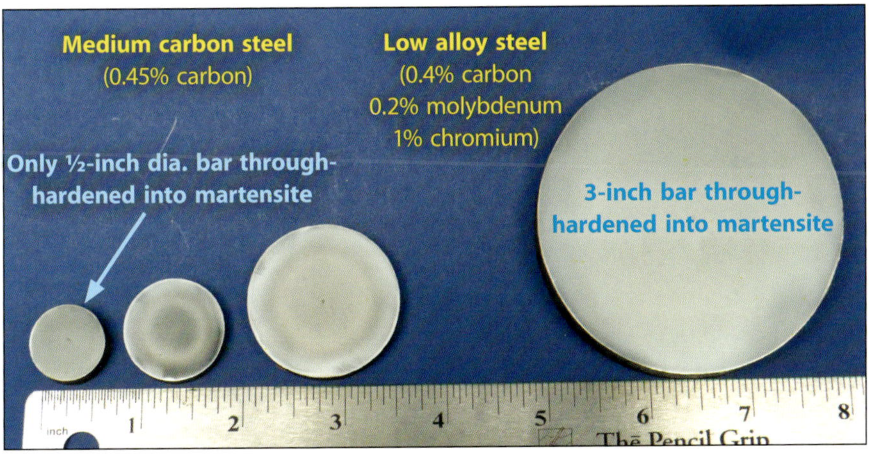

Figure 5.12: Due to its molybdenum and chromium alloying, this low alloy steel had much higher hardenability than the carbon steel. Whereas only the ½-inch diameter bar of carbon steel could be through-hardened, a bar six times that diameter of this low alloy steel could be through-hardened into martensite.

One drawback to high hardenability is that it increases the efforts needed during welding. Remember, martensite without tempering makes steel brittle. If martensite were accidentally formed during welding, it could risk the integrity of the part. Most hardenable steels are still weldable yet additional precautions need to be taken. This will be discussed further in Chapter 8.

Appendix D provides optional reading that further discusses steel hardenability and introduces diagrams used to predict the phases that form upon quenching.

5.7 Properties Established by Alloying and Heat Treatment

The priorities of those involved in designing and manufacturing often lie in how steel can be manufactured into parts while ensuring the design properties are established. Although the properties such as strength are based upon (a) steel alloying and (b) microstructure, industry rarely discusses steel in terms of its microstructure. This is largely because few facilities have the resources to examine microstructure and most professionals have no means of checking it. Instead, industry designs and manufactures steel parts using the variables they have control over: purchasing steel and processing.

So long as the alloying and processing are meticulously controlled, the right microstructure and properties will be achieved. To achieve this, steel is purchased to specific alloying requirements, possibly in a particular heat treated/processed condition and often already tested to ensure it has the intended properties (such as tensile testing). During manufacture, any process that might change the microstructure and properties of the steel is carefully controlled, such as welding and additional heat treatments.

Whether the role of microstructure is discussed or not, its silent contribution behind alloying and processing is crucial for the economic manufacture of steel parts and their safe use. Hence why it is worth learning the role of microstructure, how to manipulate it through processing (specific to steel alloying) and how that manipulation can be useful.

Table 5.2 summarizes the range of properties that the three bolt grades, heat treated in different ways, had provided. Each combination had either maximized a particular trait (such as strength or ductility) or balanced several properties. Some of these characteristics created ideal properties for the finished part, while in other cases their benefits might have helped with the manufacture of the part itself. Knowing how to select appropriate alloying and processing is what makes it possible to customize steel for applications ranging from food cans to aircraft landing gear.

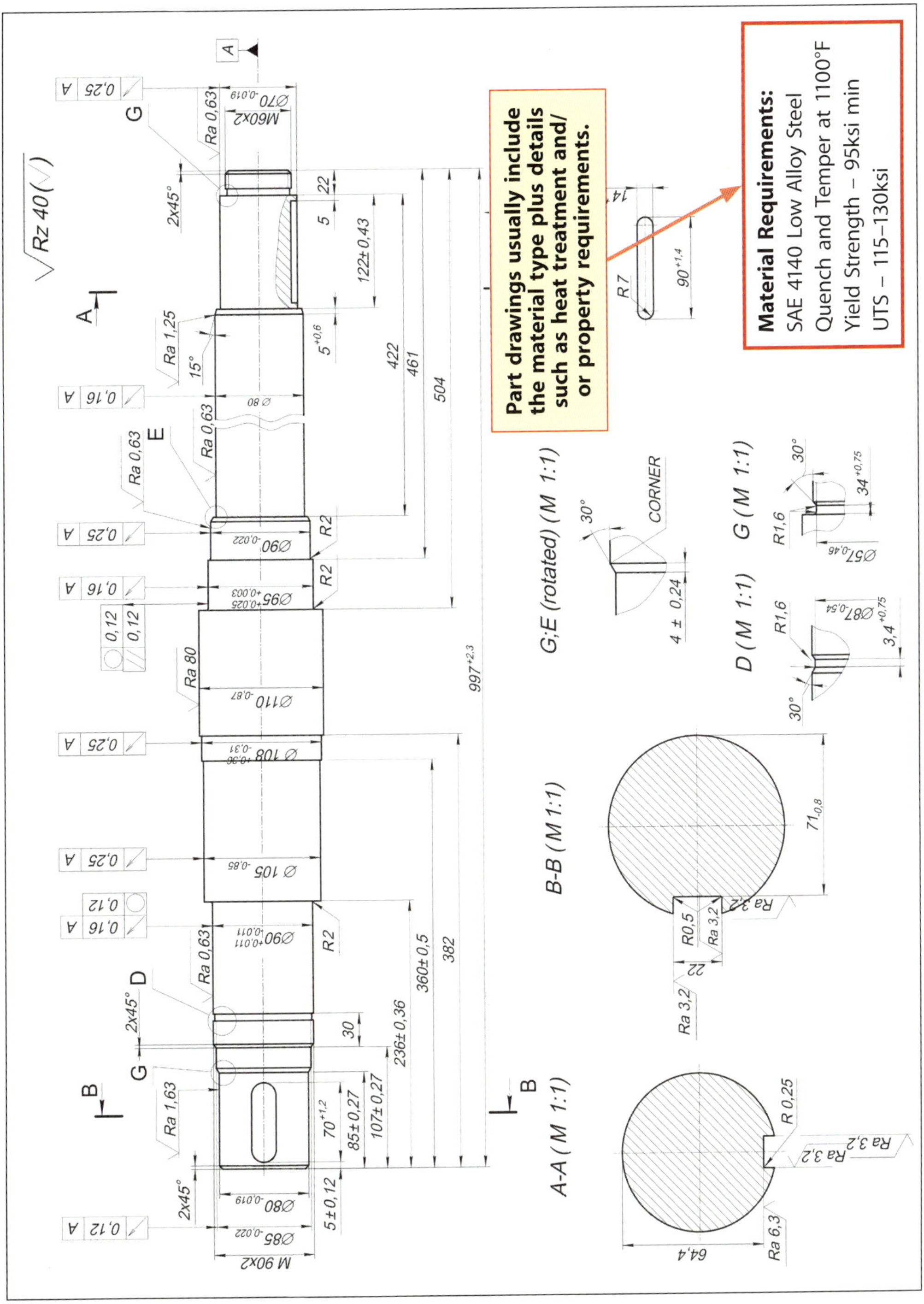
Part drawings usually include the material type plus details such as heat treatment and/ or property requirements.
Material Requirements:
SAE 4140 Low Alloy Steel
Quench and Temper at 1100°F
Yield Strength – 95ksi min
UTS – 115–130ksi
A-A (M 1:1)
B-B (M 1:1)
G;E (rotated) (M 1:1)
D (M 1:1)
G (M 1:1)
CORNER

Table 5.2: Summary of Alloying, Microstructure and Heat Treatment of the Grade 2, 5 and 9 Bolts

Bolt Grade	**Notes**	**Annealed** 1650°F then slowly furnace cooled
Grade 2 Bolt 0.15% C (Low Carbon Steel)	$0.45 per bolt Not hardenable Easy to weld	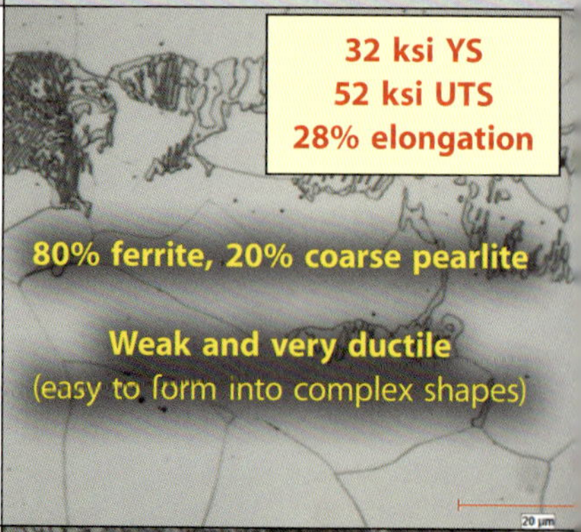
Grade 5 Bolt 0.4% C (Medium Carbon Steel)	$0.60 per bolt Thin parts hardenable Time consuming to weld	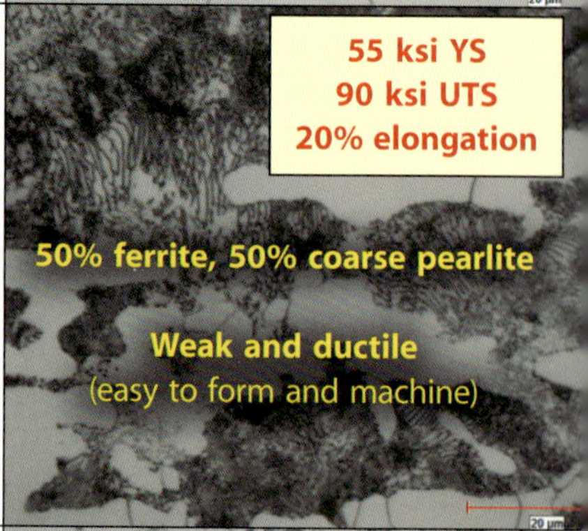
Grade 9 Bolt 0.4% C +Mo+Cr+Ni (Low Alloy Steel)	$1.50 per bolt Thick parts hardenable Time consuming to weld	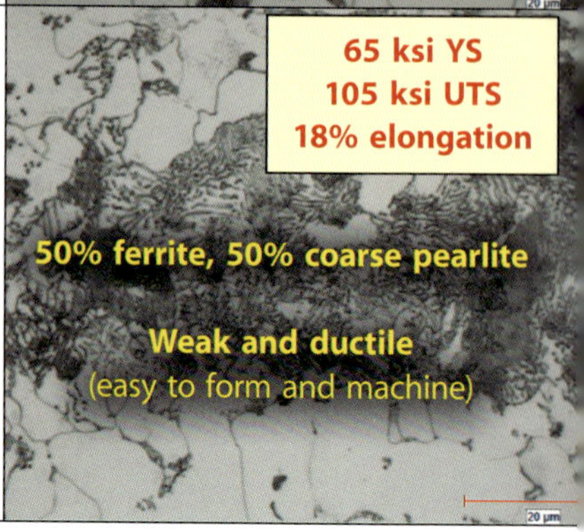

Heat Treatment	
Normalized 1650°F then air cooled	**Quenched and Tempered** 1650°F then water quenched and tempered

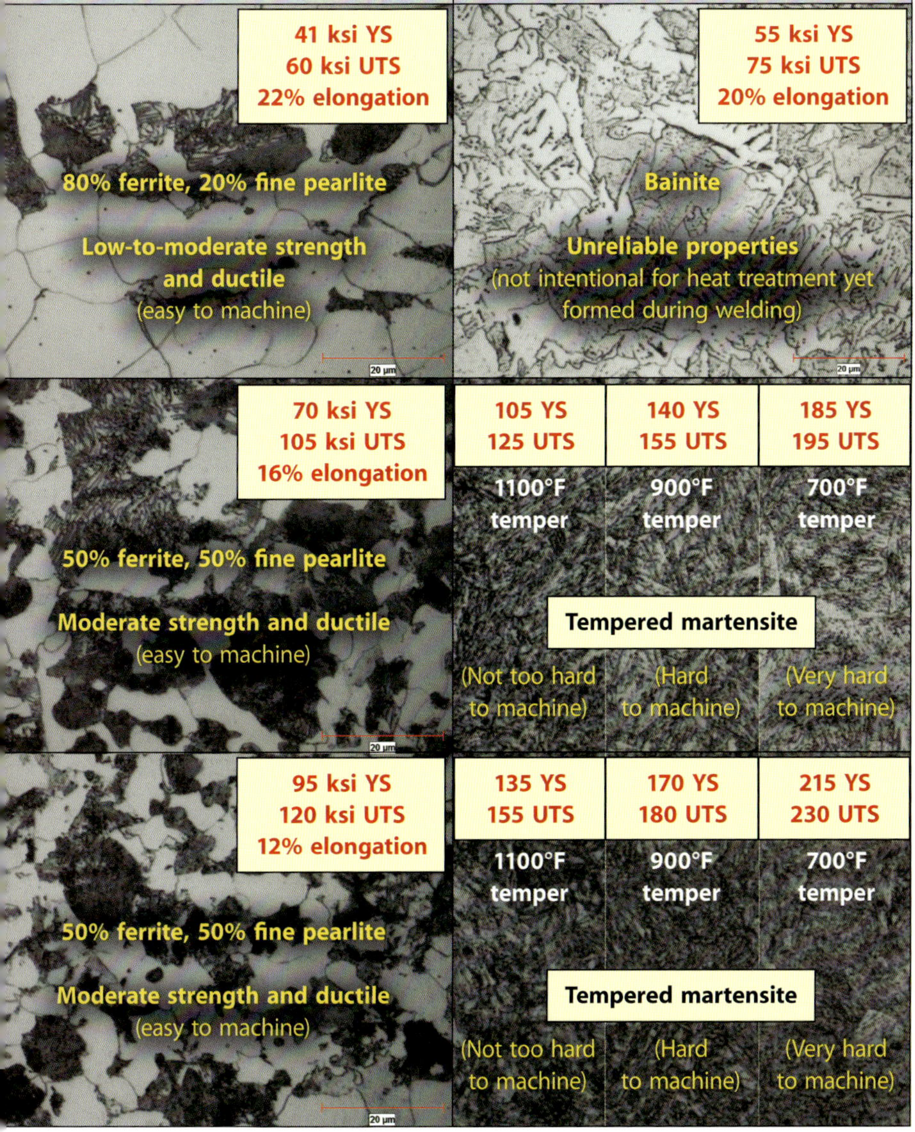

5.8 Summary of Microstructure and Heat Treatment

1. The microstructure of steel greatly affects its properties.
2. Microstructure is established through processing such as heat treatment.
3. Ferrite and pearlite:
   a. Stronger than most other metals yet not as strong as martensite.
   b. Formed during 'normalizing' by air cooling steel from red-hot temperatures. Provides a balance of moderate strength, ductility and relatively easy to work with during manufacture.
   c. 'Annealing' cools steel very slowly in a furnace, forming coarse pearlite. This sacrifices its strength to maximize both its ductility and ease to manufacture with. Once shaped, load-bearing parts are heat treated again to establish their final properties.
4. Tempered martensite:
   a. Provides high strength. Strength and ductility adjusted by tempering temperature.
   b. Formed when hardenable steels (sufficient carbon/alloying) are 'quenched and tempered'. Includes water or oil quenching from red-hot temperatures.
   c. Untempered martensite is brittle and usually unsafe.
   d. The higher the carbon and alloying, the higher the steel's hardenability and the larger the part that can be quenched into martensite.
5. Upper bainite:
   a. Not an intended phase from heat treatment yet common for less controlled processes such as welding.
   b. Formed if either (i) non-hardenable steel is quenched or (ii) hardenable steel is improperly quenched.

6. Design and manufacture cares foremost about final properties.
    a. Carefully controlling the (i) steel alloying and (ii) processing to set the microstructure is the means to establish the design properties.

The criteria of each steel 'grade' is specific enough that qualifying steel batches will behave in a similar and predictable manner.

6.0 STEEL GRADES, STANDARDS AND MATERIAL CERTIFICATES

Steel is designed, produced and sold as 'grades'. Each grade has well-defined criteria – all steel meeting those requirements can be expected to behave in essentially the same way and have predictable properties. This allows engineers to select steel and design based upon the known properties from a long list of grades. Their decision of steel grade and its condition or heat treatment is then included on the part drawing. Steel is later purchased to these same grade requirements, ensuring that the steel behaves as expected during manufacturing and later in service.

6.1 SAE Steel Grades

There are thousands of steel grades and different industries prefer different specification systems. The most basic system used in North America is overseen by the Society of Automotive Engineers (SAE). This grade series is based solely upon compositional requirements. It uses a four-digit code with the first two numbers representing the alloying family and the last two digits representing the carbon content. For example, the three bolts introduced in Chapter 2 could be classified by the SAE grades described in **Table 6.1**.

Each SAE steel grade has well-defined compositional tolerances. **Table 6.2** lists the alloying and impurity ranges permitted for SAE Grades 1015, 1040 and 8640. Steel within these tolerances will provide essentially the same behavior during processing and manufacture. For example, all steels conforming to the compositional requirements of a particular grade can be heat treated at the same temperature and be expected to achieve similar properties from that heat treatment. This means that, if processed consistently during manufacture, steel batches conforming to a specific SAE grade will repeatedly achieve the same properties.

Table 6.1: SAE Grade Designations

Grade 2 Bolt	Grade 5 Bolt	Grade 9 Bolt
0.15% carbon	0.40% carbon	0.40% carbon 0.20% molybdenum 0.45% chromium 0.45% nickel
SAE Grade		
1015	1040	8640
"10" represents carbon steel class "15" target carbon content of 0.15%	"10" represents carbon steel class "40" target carbon content of 0.40%	"86" nickel-chromium-molybdenum low alloy class "40" target carbon content of 0.40%

Table 6.2: Grade Compositional Tolerances

SAE Grade	Composition (%)							
	C	Mn	S	P	Si	Mo	Cr	Ni
1015	0.13-0.18	0.30-0.60	0.050 max	0.030 max	-	-	-	-
1040	0.37-0.44	0.60-0.90	0.050 max	0.030 max	-	-	-	-
8640	0.38-0.43	0.75-1.00	0.040 max	0.040 max	0.15-0.35	0.15-0.25	0.40-0.60	0.40-0.70

Another organization, the American Iron and Steel Institute (AISI), copies the SAE system. It is common that steel grades using the same numbering convention be referenced interchangeably between "SAE" and "AISI".

Appendix B includes additional discussion regarding some of the most common SAE steel grades in use today.

6.2 ASTM Standards

While the SAE grades have only compositional (alloying) requirements, many steels are purchased to more comprehensive standards. The most common standard series is organized by the American Society of Testing and Materials (ASTM). These standards can include (a) composition, (b) heat treatment/condition and (c) strength requirements amongst other prerequisites. This way steel can be purchased in the exact condition planned by the design engineer without requiring further processing.

For example, a common construction material for bridges and buildings is ASTM A36 carbon steel. The ASTM A36 specification states that the steel must meet the strength requirements per **Table 6.3** and its composition must fall within the **Table 6.4** limits. Having a minimum strength requirement just makes sense – it allows engineers to design using that minimum strength. Later during construction, steel can be purchased with that minimum strength. Yet equally important, the defined composition ensures that the steel behaves in a predictable manner during construction and welding.

Table 6.3: Summary of ASTM A36 Tensile Requirements[1]

Geometry	Yield Strength (ksi)	Ultimate Tensile Strength (ksi)	Elongation (%)
Plates and bars	36 minimum	58 - 80	23 min
Shapes (i.e. I-beam)			21 min

[1] Reprinted, with permission, from *ASTM A36/A36M-19 Standard Specification for Carbon Structural Steel*, copyright ASTM International, 100 Barr Harbor Drive, West Conshohocken, PA 19428. A copy of the complete standard may be obtained from ASTM International, www.astm.org.

6.3 Material Certificates

It is critical that the steel used during manufacture and construction are of the correct grades and meet the intended requirements. This is why steel purchased as bar, plate, sheet, pipe, rebar, etc. comes with a 'material certificate' or 'mill certificate'. These certificates list the relevant details of a specific batch of steel. Details listed in a material certificate usually include:

1. The steel grade and relevant specifications.
2. The steel composition.
3. The condition of the steel and applied heat treatment(s).
4. Tensile properties including yield strength, ultimate tensile strength and elongation.
5. Other details as required by the governing specification(s).

Figure 6.1 provides a mock example of a material certificate for a 2-inch diameter bar of 4140 low alloy steel.

Material certificates provide the information needed to ensure purchased steel meets the relevant specifications and design criteria. They are produced by testing a piece of steel from a large batch – meaning the certificate represents only that specific batch. They also have legal implications as the seller is declaring the steel's specific properties. Certificates are often stored for decades just in case there is a problem with the parts made from that steel batch.

6.4 Summary of Grades, Standards and Material Certificates

1. Steel is purchased by grades and/or standards. Steel that meet the criteria of a particular grade or standard should behave in a similar and predictable manner.

Table 6.4: Summary of ASTM A36 Compositional Requirements[1]

Geometry	Thickness (inches)	Composition (%)					
		C	**Mn**	**Si**	**S**	**P**	**Cu**
	≤0.75	0.25 max	-				
	0.76-1.5	0.25 max	0.80-1.20				
Plate	1.5-2.5	0.26 max	0.80-1.20	0.40 max	0.05 max	0.04 max	0.20 max
	2.5-4.0	0.27 max	0.85-1.20				
	>4.0	0.29 max	0.85-1.20				
	≤0.75	0.26 max	-				
Bar	0.76-1.5	0.27 max		0.40 max	0.05 max	0.04 max	0.20 max
	1.5-4.0	0.28 max	0.6-0.9				
	>4.0	0.29 max					
Shapes	All	0.26	-	0.4 max	0.05 max	0.04 max	0.2 max

[1] Reprinted, with permission, from *ASTM A36/A36M-19 Standard Specification for Carbon Structural Steel*, copyright ASTM International, 100 Barr Harbor Drive, West Conshohocken, PA 19428. A copy of the complete standard may be obtained from ASTM International, www.astm.org.

2. SAE grades have specific compositional (alloying) tolerances.
   a. Uses a four-digit code – the first two digits represent the steel family, the last two digits note the target carbon content.
   b. AISI copies the SAE grade system. AISI and SAE can be used interchangeably.
3. More thorough standards exist that list requirements such as composition, heat treatment and strength requirements.
   a. ASTM is a common standard series.
4. Purchased steel comes with a 'material certificate' or 'mill certificate' that list the details of that specific steel batch.

Figure 6.1: Mock Example of a Material Certificate

Steel Image Inc.
155 – 7 Innovation Drive
Dundas, Ontario
L9H 7H9, Canada

Material Certificate

Customer:	ABCD Production		
Reference:	PO# 123456		
Product Description:	2-inch round bar	**Specification(s):**	
Dimensions:	2-inch diameter 20-ft lengths		
Weight:	4,760 lbs	**Casting Heat:**	
		Condition:	

Composition (%)						
C	**Mn**	**Si**	**S**	**P**	**Mo**	
0.42	0.84	0.25	0.016	0.007	0.21	

Mechanical Properties		
Yield Strength (ksi)	**Ultimate Tensile Strength** (ksi)	
106.3	130.1	

Notes:	Heat treated at 1550°F (843°C) for 120 minutes, oil quenched. Tempered at 1150°F (621°C) for 180 minutes. We certify that the data listed is correct and in compliance with the referenced specification(s).

		November 4, 2022
	AISI / SAE 4140	
	98765	
	Quenched and Tempered	
Cr	**Ni**	**Cu**
0.99	0.16	0.32

Elongation (%)
24
Quality Manager

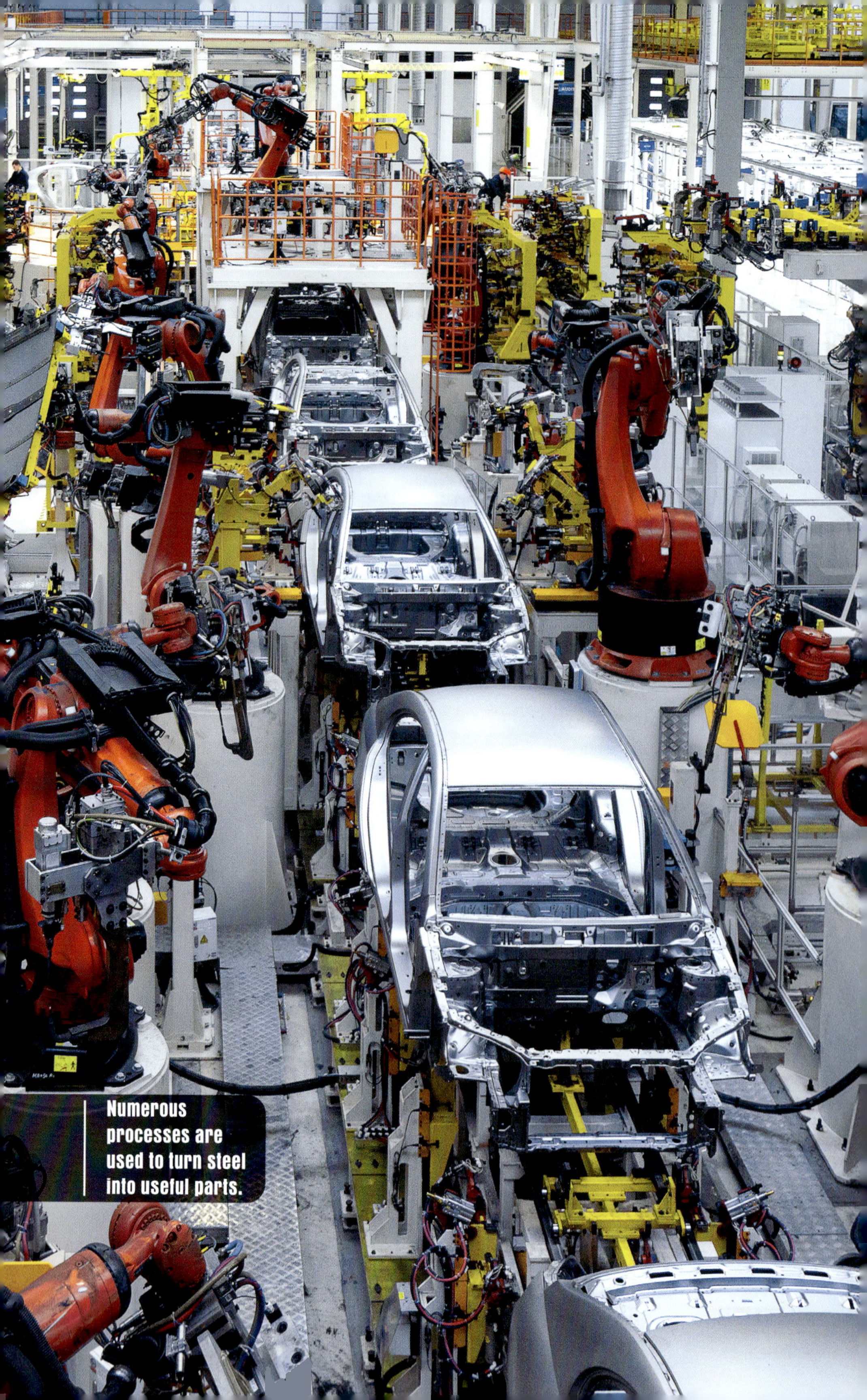

Numerous processes are used to turn steel into useful parts.

7.0 SHAPING STEEL AND MAKING PARTS

Now that the key fundamentals of steel have been introduced, we can begin to appreciate how they must be considered during the manufacture of useful components. There are a variety of methods used to transform steel into finished products, each having different benefits and nuances. These methods affect the steel in various ways – each requiring consideration to ensure both the correct geometries and steel properties are achieved.

Throughout this book, bolts have been used to demonstrate the fundamental principles of steel behavior. However, there are innumerable components made of steel. As we discuss the processes used to shape steel into different parts, we will introduce several other steel components and summarize how they are made.

7.1 Casting

Some components are made by pouring molten steel into shaped molds, forming a 'cast' part (**Figure 7.1**). One advantage of casting is that it produces near-shape geometries. For example, the irregular geometries of train suspension components and valve bodies can be cast to their near-final shapes (**Figures 7.2** and **7.3**). Once cooled and removed from their molds, castings are often heat treated by either normalizing or quenching and tempering. Since the dimensions of castings are not precise, any critical surfaces or features that need to be dimensionally accurate are then machined.

Figure 7.1: Some components are cast to their near-final shape. Molten steel is poured into molds designed for specific shapes. After the castings have cooled, they are removed from their molds, heat treated and critical surfaces/features are machined.

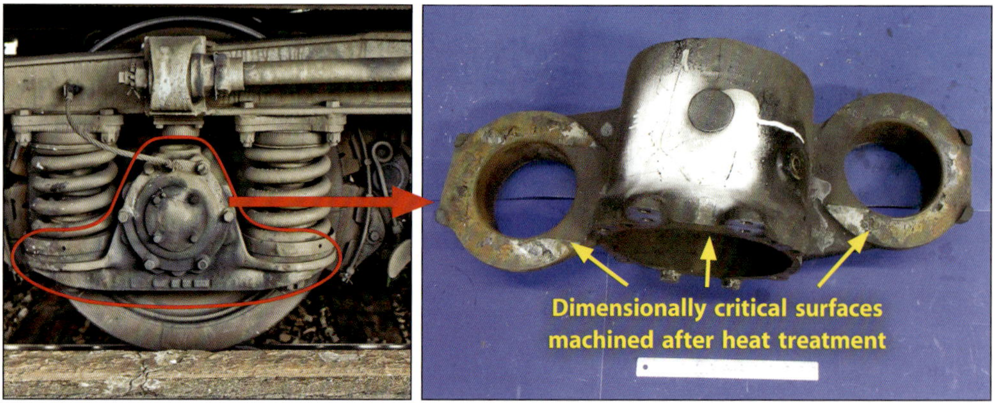

Figure 7.2: Journal boxes on train car suspensions are cast. This journal box comprised of medium carbon steel (0.35% C, 1.85% Mn) cast to shape. After cooling and being removed from its mold, this casting was quenched and tempered to its intended strength. Dimensionally critical surfaces, such as its bore and journal faces, were then machined.

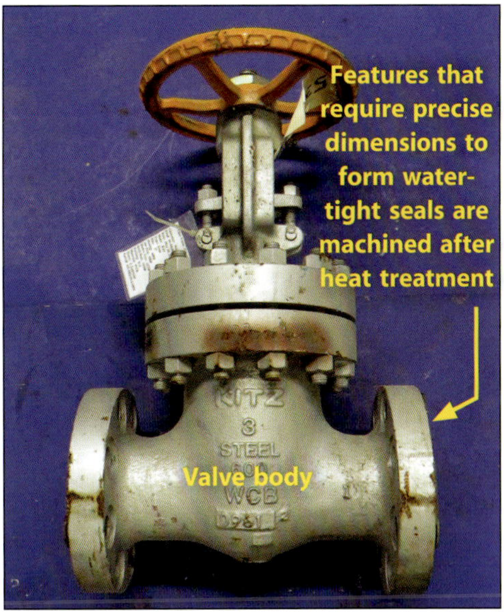

Figure 7.3: Medium and large valve bodies are often cast. This 3-inch ASTM A105N valve body had been cast using low carbon steel (0.25% C) and then normalized. Surfaces critical for maintaining water-tight seals were machined last.

7.2 Hot Forming

Most raw steel available for purchase has been hot formed into common shapes such as sheet, bar or tube. These products are produced by heating steel blocks and running them through mills that squeeze, crush and form them into shape (**Figures 7.4** and **7.5**). If no other heat treatment is applied, these products are described as 'hot formed' or 'hot rolled' steel. They are sold to manufacturers who then further shape, machine and/or weld them into their final creations. Many building materials such as I-beams and rebar are used in the 'hot formed' condition (**Figure 7.6**).

Steel shaped by hot forming may also be sold in a heat treated condition – annealed, normalized or quenched and tempered. Many facilities prefer to purchase steel in either the condition easiest to manufacture with or in the final condition they intend to use it. For an example of the latter, shafts are often made from bar purchased in the normalized or quenched and tempered conditions, requiring only machining to produce a shaft (**Figure 7.7**).

Figure 7.4: Most steel available for purchase has been hot formed into common shapes such as bar, plate, sheet, tube, I-beam, etc. The majority of hot formed steel is air cooled which produces ferrite and pearlite microstructures. These stock materials can then be further heat treated, machined, formed and/or welded into assemblies.

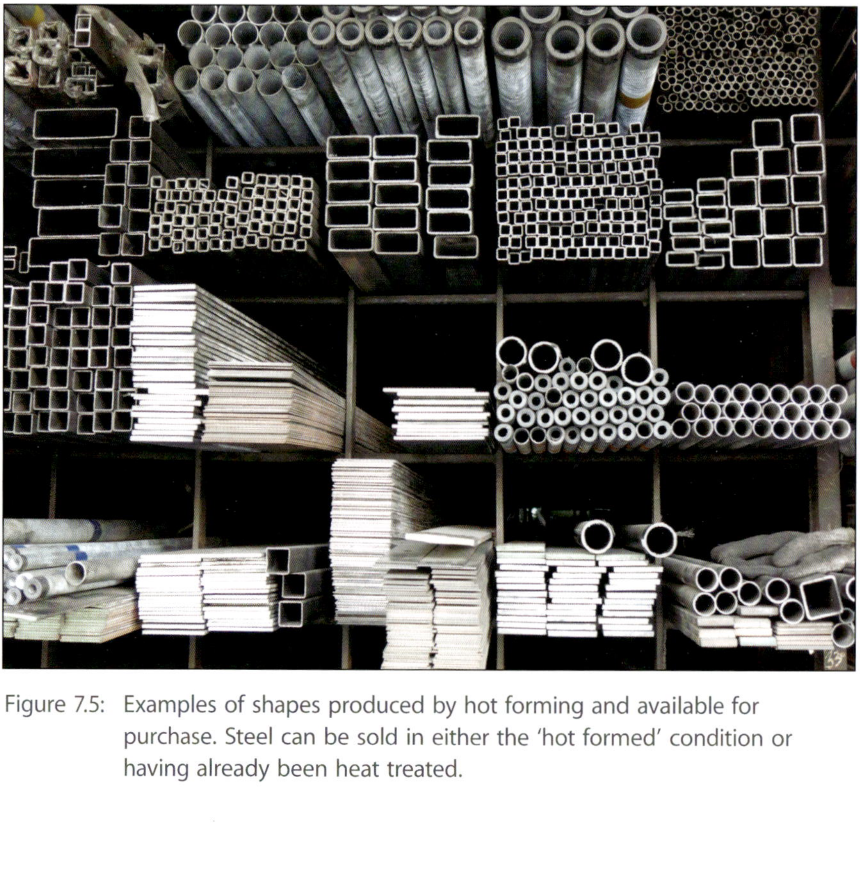

Figure 7.5: Examples of shapes produced by hot forming and available for purchase. Steel can be sold in either the 'hot formed' condition or having already been heat treated.

Figure 7.6: I-beams used for buildings and bridges usually comprise of low carbon steel such as ASTM A36 (<0.29% C) in the hot rolled condition. Their low carbon content makes them easy to weld, needed for quick and economical construction.

Figure 7.7: Shafts are used in everything from jet engines, freighters, cars, trains to kitchen blenders. Therefore, the material and strength requirements vary quite a bit. Still, most shafts are machined from bars of medium carbon (>0.30% C) or low alloy steels. Common alloys include SAE 1045 carbon steel, SAE 4140 and 4340 low alloy steel. Bars are usually purchased already normalized or quenched and tempered, requiring only machining to produce the shaft.

7.3 Forging

While steel can be incredibly strong at safe-to-touch temperatures, it is soft and malleable at red-to-white-hot temperatures. At these extremely hot temperatures, even the strongest steels can be hammered and forged to shape. This is not a recent discovery – this technique has been used by blacksmiths for thousands of years. The biggest difference from ancient blacksmithing to modern forging is that we now use hydraulic forging presses (**Figure 7.8**). Either a person manually adjusts the hot steel beneath a forging press or steel is battered into shaped molds.

Figure 7.8: Forging is performed by heating steel to nearly white-hot temperatures, often in the range of 2150 to 2350°F (1177-1288°C) and hammering it to shape. Whereas blacksmithing included manual labor, modern forging is completed using presses.

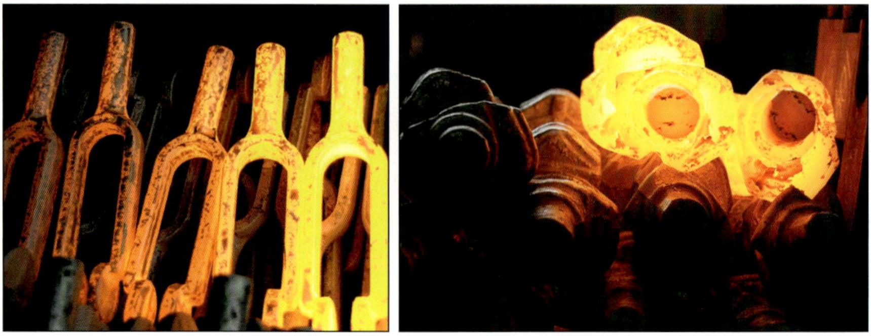

Figure 7.9: Examples of shapes that forging can produce. Forging is completed at such extreme temperatures that, when air cooled afterward, they can form less-than-ideal microstructures and poor properties. Hence why forgings are often heat treated at a later time.

Forging allows for the shaping of large and/or strong steel into essentially any shape (**Figure 7.9**). Once shaped and air cooled, forgings are often normalized or quenched and tempered to ensure the intended properties are achieved. Since forgings can only get close to the final shape, features that require exact dimensions must be machined. **Figures 7.10** and **7.11** summarize how forging is used to produce crane hooks and airplane landing gear.

Figure 7.10: Crane hooks are usually forged. This hook comprised of high strength SAE 4340 low alloy steel (0.4% C +Mo+Cr+Ni). It had been forged to shape, cooled, then quenched and tempered to achieve the design strength. The hook shank was then machined to fit into the swivel joint.

Figure 7.11: To endure hard touchdowns, landing gear components are made of high strength steel such as SAE 4340 low alloy steel (0.4% C +Mo+Cr+Ni). The major components are forged to near-final shape. Long after cooling down, the forgings are quenched and tempered before being machined to their final shapes.

7.4 Cold Forming

Forging at nearly white-hot temperatures allows for the shaping of large steel parts. It is also possible to form and shape thin ductile steel without heating. Cold forming is the bending or shaping of sheet, bar, tube, etc. into a particular geometry at room temperature. Usually just referred to as 'forming', there are several forming subcategories including stamping, drawing, bending and pressing. Most methods use dies or custom tooling to shape steel into specific shapes and contours. The steel being formed must have adequate ductility to flow into the intended shape. One advantage over forging is that cold forming can produce smooth surface finishes useful for decorative applications, sanitary purposes and are easy to coat (**Figure 7.12**).

Figure 7.12: Forming is used to shape thin steel such as food cans and car bodies. Unlike forgings, cold formed parts can have smooth and clean surface finishes. This can provide an aesthetically pleasing finish once coated.

7.5 Cold Working and Cold Rolling

If steel is deformed at safe-to-touch temperatures, the atoms and microstructure within the steel distort to accommodate the shape change. This atomic and microstructure movement 'cold works' the steel which increases its yield strength. If compressive/crushing loading is also included, cold working can also increase its ultimate tensile strength.

'Cold rolling' takes advantage of this; squeezing steel between powerful rolls to reduce its thickness. The cold working it causes increases the steel strength at the loss of its ductility. **Figure 7.13** illustrates how cold rolling reduced the thickness of low carbon steel (0.15% C) and cold worked its microstructure. **Table 7.1** summarizes how this affected the steel's strength and ductility.

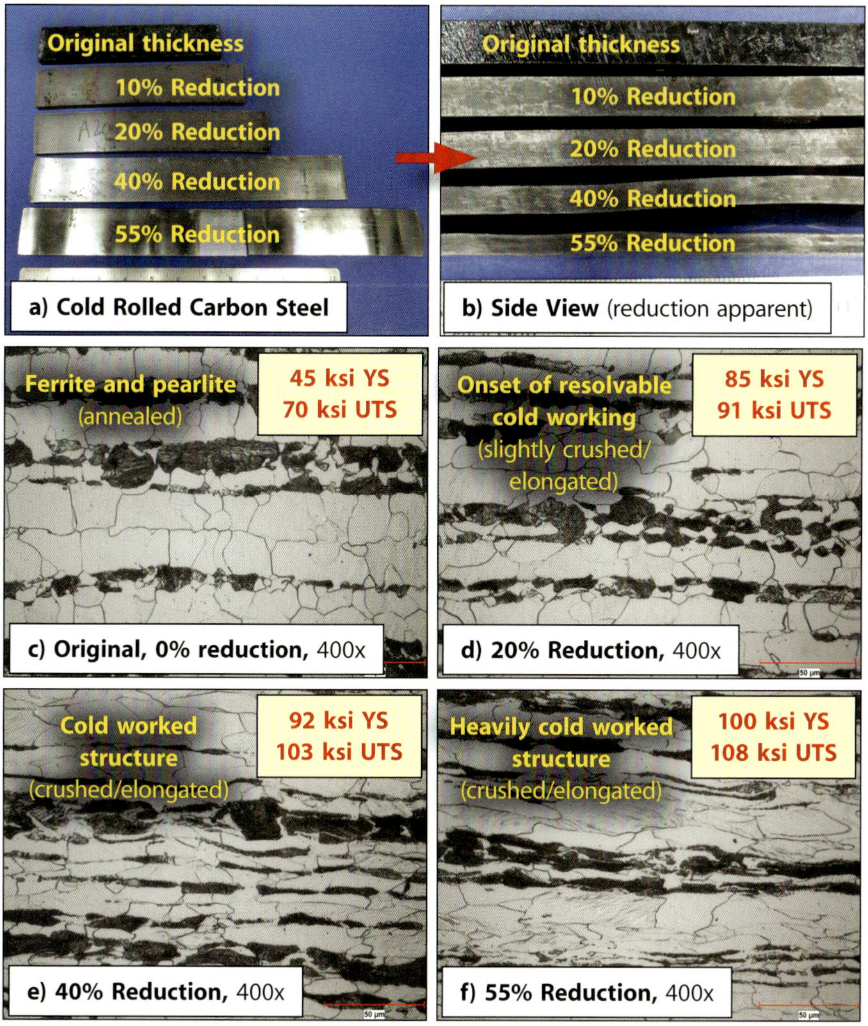

Figure 7.13: The cold rolling of a low carbon steel (0.15% C) reduced its thickness and crushed the microstructure. The greater the reduction, the more cold worked the microstructure became and the greater the increases in its yield strength and UTS.

Table 7.1: Effects of Cold Rolling on Annealed Low Carbon Steel (0.15% C)

Cold Rolled Reduction	Yield Strength (ksi)	UTS (ksi)	Elongation (%)
Original Thickness	45	70	36
10%	78	82	19
20%	85	91	14
40%	92	103	10
55%	100	108	8

Cold rolled and cold formed steels are used for applications that require high strength and benefit from the smooth surface finish it produces – such as springs. There are several common methods of making springs. Large springs are usually made by hot-forming hardenable steel and then quenching and tempering it. Medium-sized springs are usually made by cold forming annealed, hardenable steel before quenching and tempering. Yet the easiest method, used for making small springs, is to simply twist high carbon, cold formed steel wire into the coil shape (**Figure 7.14**).

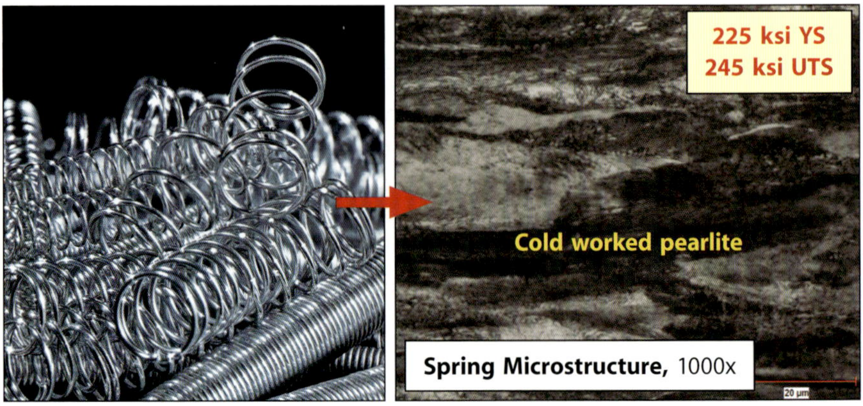

Figure 7.14: Springs require very high yield strengths to ensure they always spring back to shape. One method of manufacturing small springs is to use cold formed SAE 1080 high carbon steel (0.8% C) wire. The wire is simply twisted into shape.

7.6 Machining

Whereas casting, forging and forming can shape parts to approximate dimensions, machining can produce features with precise geometries. Machining is the removal of material to fabricate an intended shape. Machining can create complex shapes from purchased steel or touch-up parts already shaped by other means (**Figure 7.15**).

Figure 7.15: Machining is essential for the production of many parts. Skilled machinists or computer-controlled equipment can machine steel into complex geometries with dimensional accuracy.

There are many machining processes including cutting, drilling, milling and turning on a lathe. Some are manually operated by skilled machinists while others are computer automated. Most machining equipment works on the principle of something turning while something sharp cuts the steel – the differences between equipment lie in what turns and what cuts. **Figure 7.16** illustrates how a lathe turns an entire part while a stationary cutting bit shaves off metal. **Figure 7.17** shows how a milling machine rotates a cutting tool that trims a fixed, stationary part. To prevent overheating, water-based coolants are often poured onto the steel while machining.

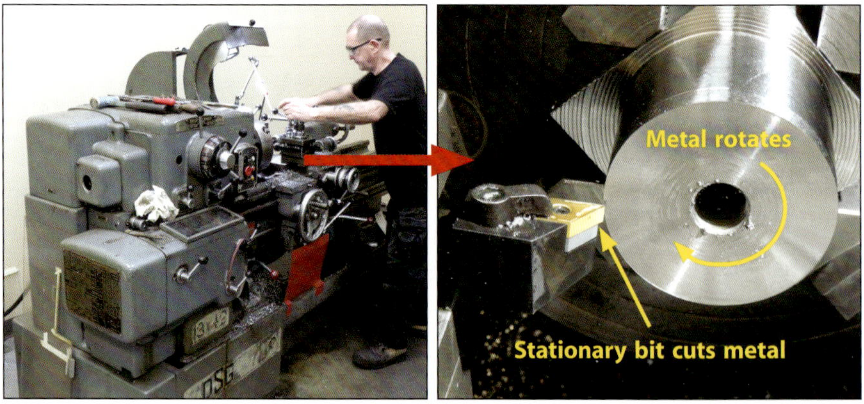

Figure 7.16: A lathe rotates the steel and uses a stationary cutting tool to shave off metal. Water-based coolant is applied to prevent the metal piece from overheating and changing the microstructure.

Figure 7.17: A milling machine holds the part stationary while a high-speed rotating tool cuts it to shape. Water-based coolant is usually applied to prevent the metal from overheating.

7.7 Surface Hardening

Recall that if a part were quenched and left as untempered martensite, the part would be brittle and unsafe for use. Yet if only the surface layer comprised of untempered martensite while the material beneath had good ductility, the part would not suffer the same embrittling effect. For parts that require high surface strength or hardness, processes called 'surface hardening' are used to exploit this phenomenon. The two most common methods of surface hardening are (a) carburizing and (b) induction hardening.

'Carburizing' diffuses carbon into a part's surface at red-hot temperatures. Carburizing is completed within specialized furnaces filled with carbon-rich gases that infuse carbon into the steel's surface. Upon quenching, the high carbon content along the surface produces a hard surface layer. **Figure 7.18** illustrates the surface microstructure of a gear that had been carburized. This combination of a hard surface with a softer core provides high surface strength without embrittling the part.

'Induction hardening' is another common surface hardening method. While conventional furnaces slowly heat an entire part, induction hardening uses magnetic fields that almost instantly heats a part's surface. This allows for the surface to be independently heated and quenched into martensite, leaving the core microstructure unchanged. A light temper might be applied afterward to marginally reduce the surface hardness. **Figure 7.19** summarizes how induction hardening is used as an alternative method for surface hardening gears.

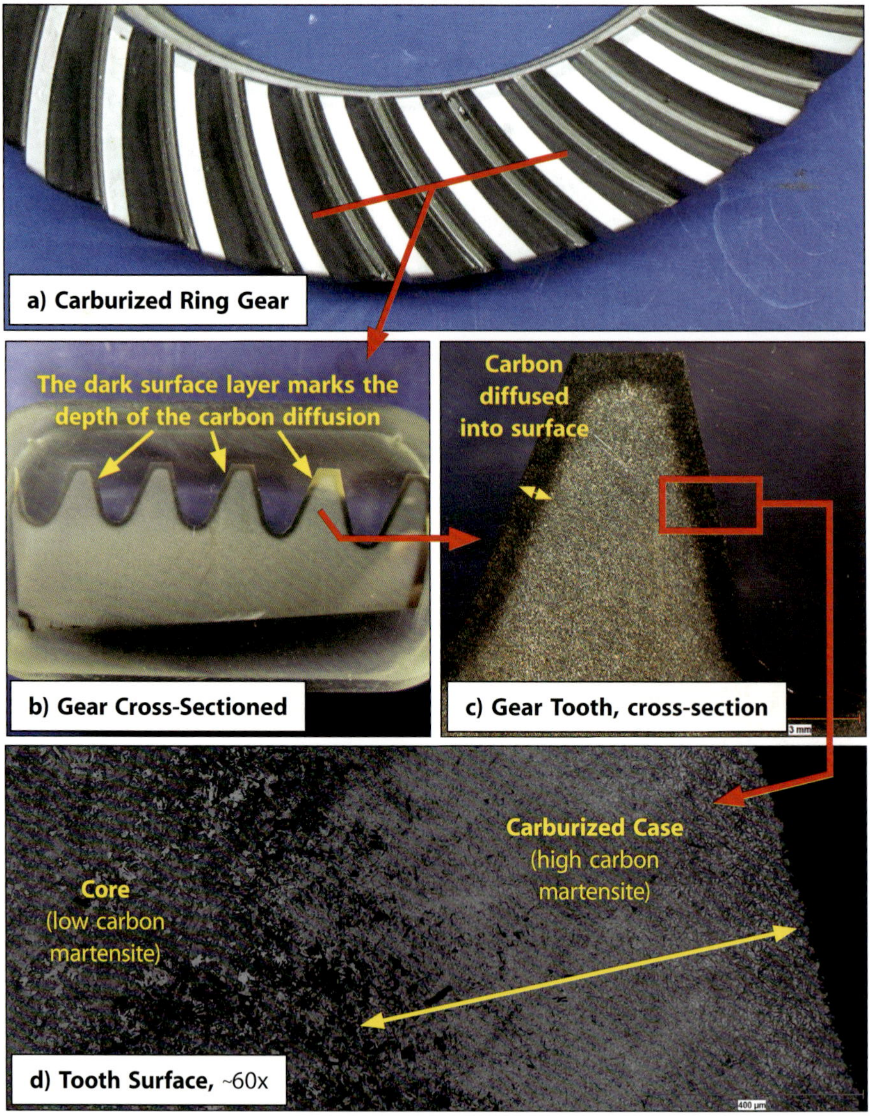

Figure 7.18: Many gears are made by carburizing SAE 8620 low alloy steel (0.2% C +Mo+Cr+Ni). After machining to near-shape, carburizing diffuses carbon into the surface. Whereas the core has only 0.2% carbon, the surface reaches close to 0.8% carbon. Upon quenching, the surface forms a layer of hard martensite. Although SAE 8620 has enough Mo+Cr+Ni alloying for the core to also form martensite, it remains softer and ductile because of its lower carbon. The combination of the hard/strong surface and ductile/tough core provides ideal properties for gears. The gear will then undergo a final machining (grinding) operation to ensure dimensional accuracy.

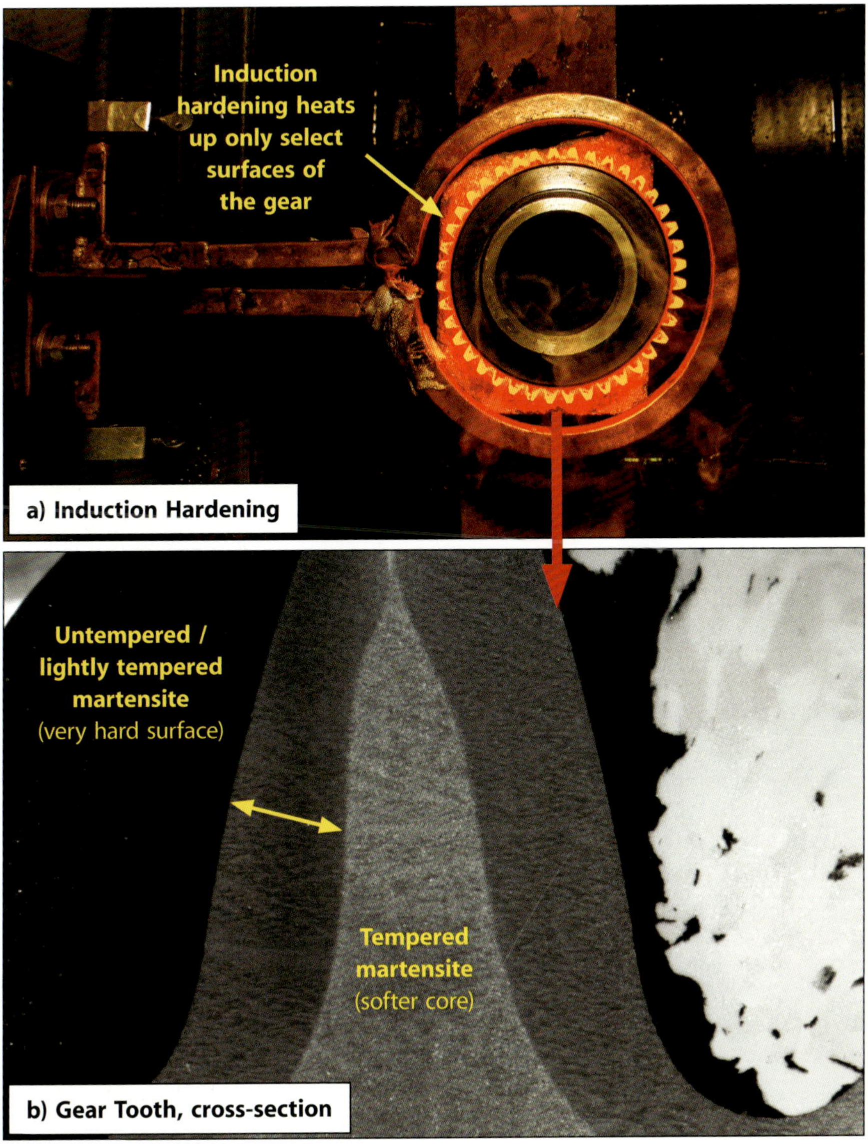

Figure 7.19: A second method of producing gears is by induction hardening low alloy steels such as SAE 4140 (0.4% C +Mo+Cr). The steel is first quenched and tempered to a moderate strength. After machining to shape, the teeth are induction hardened. Induction hardening heats the teeth surfaces to red-hot temperatures in mere seconds. Upon quenching, only the heated surface transforms into fresh martensite. A light temper might be applied afterwards. The combination of lightly-tempered martensite at the surface and well-tempered martensite in the core provides high surface strength without embrittlement.

7.8 Summary of Shaping Steel and Making Parts

1. Molten steel can be cast to near-shape in molds.
2. Most steel available for purchase has been hot formed into simple shapes such as sheet, bar or tube.
3. Steel can be forged and rolled into shapes at red-to-white-hot temperatures.
4. Castings, forgings and hot formed steel may require heat treatment, such as normalizing or quenching and tempering, to achieve their design properties.
5. Small or thin steel parts can be cold formed at ambient, safe-to-touch temperatures.
6. Cold forming and deformation at ambient temperatures introduce cold working to the microstructure that changes the steel properties. Cold rolling uses this to increase the steel strength.
7. Surface hardening can create very hard and strong surfaces. The combination of a hard surface with a softer core provides high surface strength without embrittling the part.
8. Machining removes material to produce shaped parts or to touch-up dimensionally critical features of parts already shaped by other means.

Welding joins steel together. It removes the size barrier of what we can build, allowing us to construct everything from pipelines that span continents to the massive freighters that sail between.

8.0 WELDING

Welding is used to fuse and join metal pieces together. It is an essential process for the manufacture, construction and repair of most steel assemblies. Welding is used to erect the buildings, towers and bridges that define our cities. It is critical for the production of common goods such as cars and appliances. It is also needed to build and maintain the power plants, dams, wind turbines and refineries that provide power to our civilization. Without welding, our world would look quite different.

8.1 Two Common Welding Methods

Perhaps the most quintessential welding method is called 'stick' welding. **Figure 8.1** illustrates how two sheets of low carbon steel (0.15% C) were welded together by this method. Stick welding uses arcing electricity to melt the steel within 'electrodes', fusing the steel pieces together once it solidifies. **Figure 8.2** displays several electrodes which comprise of steel rods surrounded by flux. As an electrode melts, the flux emits a shielding gas that prevents air from contacting the molten metal – air mixing with the molten steel would contaminate and ruin the weld.

Another common welding technique is 'metal inert gas' (MIG) welding. Instead of using short electrodes covered with flux, it feeds wire from a roll and releases a shielding gas from a cylinder tank (**Figure 8.3**). During welding, an electric arc melts the wire being fed into the weld zone while under the protection of the shielding gas. Although MIG equipment is less mobile than stick, it allows for a faster rate of welding.

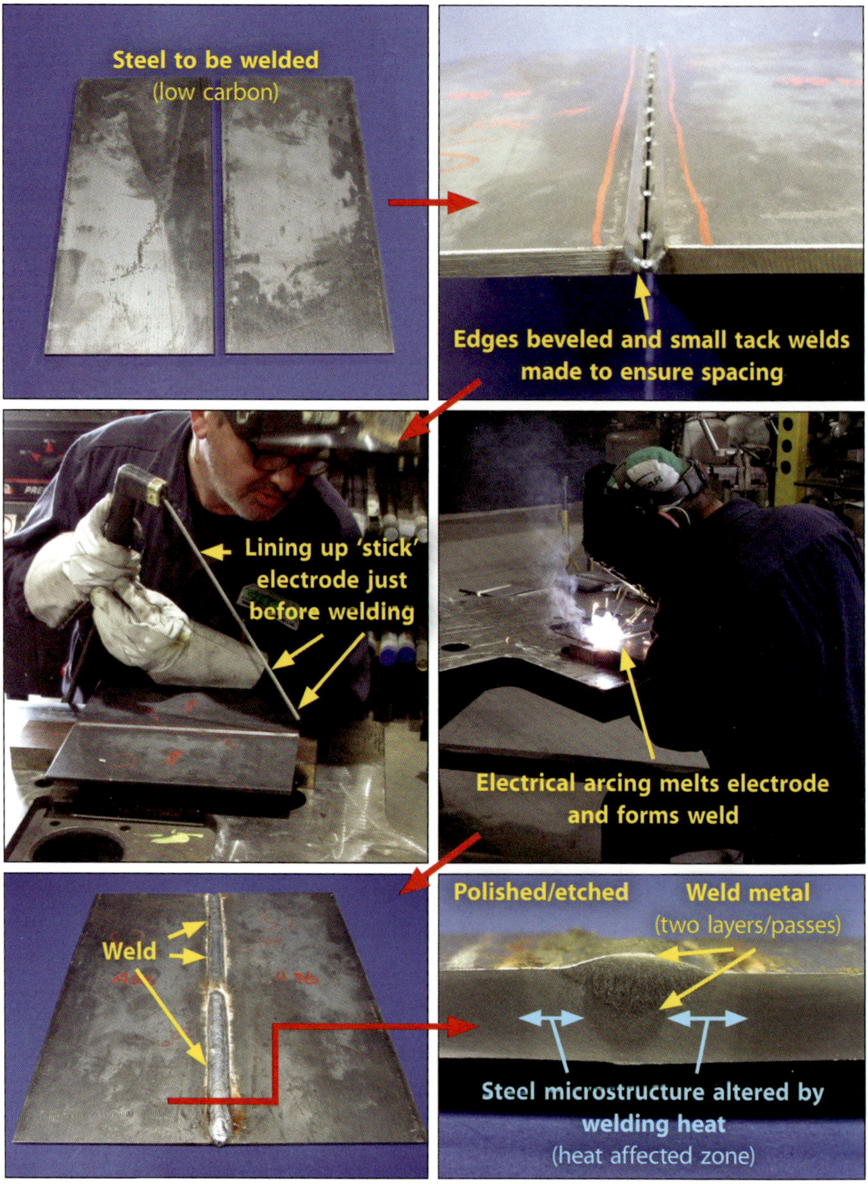

Figure 8.1: Two sheets of low carbon steel (0.15% C) were welded together by 'stick' welding. The edges of the sheets were beveled to form a 'V', lined up together and then welded in two passes. Stick welding creates an electric arc that melts the steel within an 'electrode'. Once the melted steel resolidified, the added metal fused the two pieces together.

Figure 8.2: Stick 'electrodes' comprise of steel rod surrounded by flux. During welding, electrodes are held by an 'electrode holder' and high current passes through the rod, creating an arc at its tip. The arc melts both the steel rod and the flux. The melting flux emits a gas that shields the molten metal from air. The welding sequence illustrated in Figure 8.1 was produced by this method.

Figure 8.3: MIG welding equipment comprises of a gun that feeds wire and creates an electric arc to melt the wire. A shielding gas from a cylinder tank is simultaneously released at the weld zone, displacing and preventing air from contacting the weld metal while it is molten.

There is a long list of other welding methods that use heat sources ranging from lasers, friction, electron beams to even explosives. If you understand the principles of stick and MIG welding, it sets the foundation for understanding other welding methods. This requires considering both (a) the added/melted metal and (b) how the heat during welding affects the surrounding base metal microstructure.

8.2 Weld Metal Strength

The weld size and its strength should marginally exceed the steel pieces being welded. This way the weld avoids becoming a weak link in the assembly. If a single welding pass does not produce a large enough weld, layers can be welded overtop of one another until the weld is of sufficient size. The weld made in **Figure 8.1** comprised of two welding passes.

Weld filler metals (electrodes and wire) are rated to the ultimate tensile strength (UTS) they are guaranteed to achieve. Since welding is completed in a wide range of conditions (hot summer days, cold winter days, overtop of other hot welds, etc.), the weld microstructure cannot be readily controlled or planned. Therefore, without being able to guarantee what microstructure is formed, the only way to ensure the weld joint is strong enough is by overcompensating with alloying – adding enough alloying that it will always be strong enough regardless of what microstructure is formed. Phrased differently, the electrodes and weld wire purchased will have enough alloying that they will always meet their minimum strength requirement.

Welding wire and electrodes are always low carbon so they do not form brittle martensite. Common alloying includes manganese, molybdenum, chromium, vanadium, titanium and/or niobium. **Table 8.1** lists two common weld metals/electrodes used for MIG and stick welding of low carbon steel. Both have a minimum 70 ksi UTS requirement achieved predominantly through manganese alloying.

Table 8.1: Examples of Welding Wire and Electrode Alloying

Welding Method	Stick	MIG
Wire/Electrode Designation	7018	ER70-3
Minimum UTS	70 ksi minimum	
Primary Alloying	<0.15% C <1.6% Mn Low levels of Ni, Cr, Mo, V permitted	0.06 – 0.15% C 0.9 – 1.4% Mn

8.3 Weld Heat Affected Zone (HAZ)

The electrical arcing that melts the electrode or wire also heats the steel piece(s) being welded. This melts some of the steel and the high heat changes the adjacent microstructure. **Figure 8.4** displays the 'heat affected zone' of the low carbon steel (0.15% C) welded in **Figure 8.1**. After welding, the surrounding cold steel had rapidly cooled the weld zone, effectively quenching the weld region and forming upper bainite. Upper bainite is common for heat affected zones of welded low carbon steel.

It is this quenching effect from the surrounding cool steel that can present an issue. There is no concern for low carbon steels which are not hardenable and do NOT form martensite when quenched. For higher carbon hardenable steels, brittle martensite could form and the weld zone might crack – both making the weld unsafe. This brittle effect had previously been demonstrated by welding a Grade 9 low alloy steel bolt in a similar manner as a Grade 2 low carbon steel bolt (**Figure 8.5**). When impacted, the Grade 2 bolt remained ductile and bent while the Grade 9 bolt snapped at the edge of its weld.

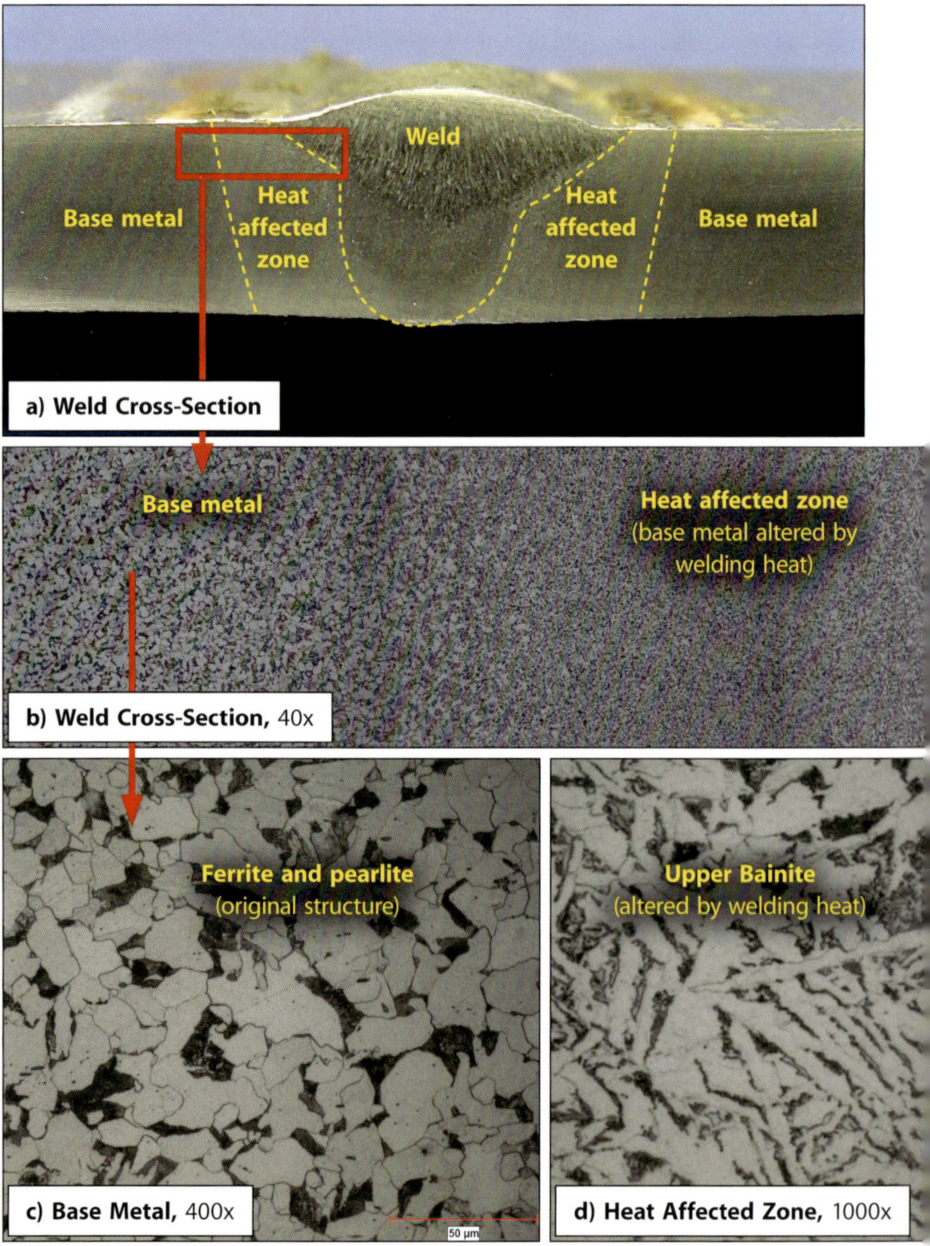
Weld
Base metal
Heat affected zone
Heat affected zone
Base metal
a) Weld Cross-Section
Base metal
Heat affected zone
(base metal altered by welding heat)
b) Weld Cross-Section, 40x
Ferrite and pearlite
(original structure)
c) Base Metal, 400x
50 µm
Upper Bainite
(altered by welding heat)
d) Heat Affected Zone, 1000x

Figure 8.4: As well as melting and adding metal, the welding heat changes the base metal microstructure around the weld. The original structure of this low carbon (0.15% C) steel had been ferrite and pearlite. The 'heat affected zone' had become hot enough and then cooled fast enough to form upper bainite. Upper bainite is common and amongst the preferred phases to form in a weld heat affected zone.

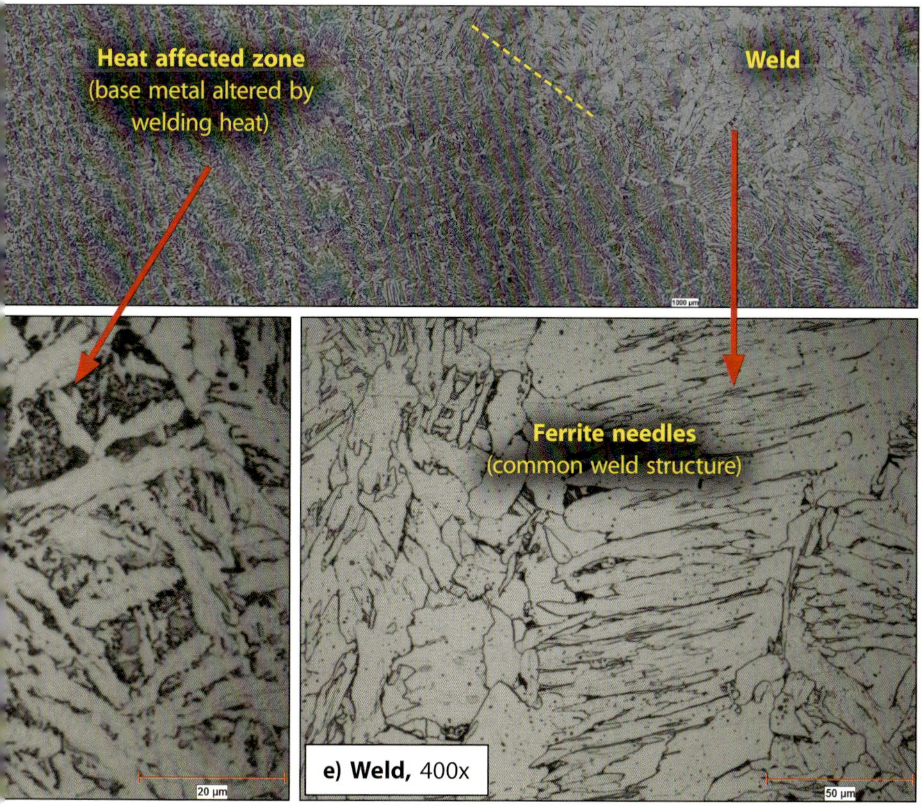

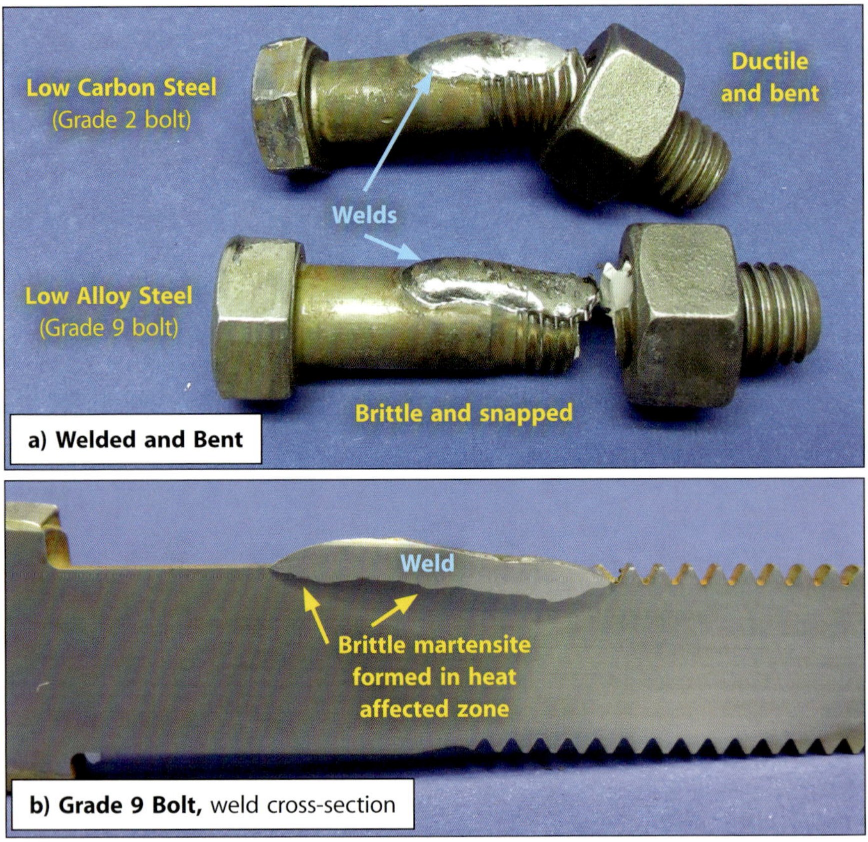

Figure 8.5: Without preheating before welding on a hardenable Grade 9 bolt (0.4% C +Mo+Cr+Ni), the weld zone quenched fast enough to form brittle martensite. As a result, the Grade 9 bolt snapped when impacted. In contrast, the low carbon Grade 2 bolt (0.15% C) still had ample ductility within the weld zone because it had not formed martensite.

Hardenable steels can be welded – they just require additional steps. Stick or MIG welding of hardenable steel requires using at least one of two strategies: (1) heating the surrounding steel before welding to prevent the quenching effect and/or (2) heating the weld afterward to temper any martensite that had formed. Preheating before welding will help prevent or at least minimize the formation of martensite. Post-weld heating afterward will temper and return ductility to any martensite that had formed. **Figure 8.6** illustrates how SAE 4140 low alloy steel (0.4% C +Mo+Cr) sheets were preheated to 500°F (260°C) before welding them together.

Figure 8.6: Hardenable steels can be welded by preheating the surrounding steel. In this case, SAE 4140 low alloy steel (0.4% C +Mo+Cr) was preheated to 500°F and then welded. Preheating slowed the rate of cooling, avoiding the formation of brittle martensite and cracking. Larger steel parts may also require post-weld heat treating to temper any martensite that might form.

There is one more thing to consider when welding hardenable steels in the quenched and tempered condition; if the heat applied before or after welding is hotter than its original tempering temperature, that heat will further temper and weaken the base steel. That is why stick or MIG welding should only be completed on hardenable steels that have been tempered at temperatures hotter than the preheating or post-weld heating needed. If its original heat treatment is not known, hardness testing can be used to approximate its tempering temperature.

Ultimately, the procedures used for welding low carbon and hardenable steels are quite different. The reason why many large constructions such as bridges, buildings and ships are made with low carbon steel is that they are NOT hardenable, making them both easy and relatively fast to weld. Low carbon steel equipment can also be readily weld repaired if they become cracked or damaged. In contrast, higher carbon, hardenable steels require additional steps that complicate and slow welding. If steps are missed, it could make the weld joint unsafe. Ultimately, whether manufacturing or repairing, the steel grade(s) being welded must be known to determine the best practices. If the steel is hardenable, its heat treatment or hardness must also be known.

8.4 Summary of Welding

1. The weld size and strength should marginally exceed that of the steel pieces being welded.

2. Weld electrodes and wire are purchased based upon the minimum ultimate tensile strength (UTS) they are guaranteed to obtain.

    a. Weld strength is achieved by overcompensating with alloying. Sufficient alloying ensures adequate strength no matter what microstructure is formed.

3. The heat of welding affects the adjacent microstructure of the steel being welded.

    a. Low carbon, non-hardenable steels are easy to weld because they do NOT form martensite during welding. This is why low carbon steels are often preferred for constructions requiring lots of welding.

    b. Hardenable steels can form brittle martensite, making their welds dangerous if additional steps are not taken during welding. These additional steps can include heating the steel before and/or after welding.

4. To develop the best practices to weld a part, the steel grade(s) must be considered. If the steel is hardenable, its heat treatment or hardness must also be known.

Water is the relentless enemy of steel.

9.0 WATER CORROSION AND PREVENTING IT

9.1 Water Corrodes Steel

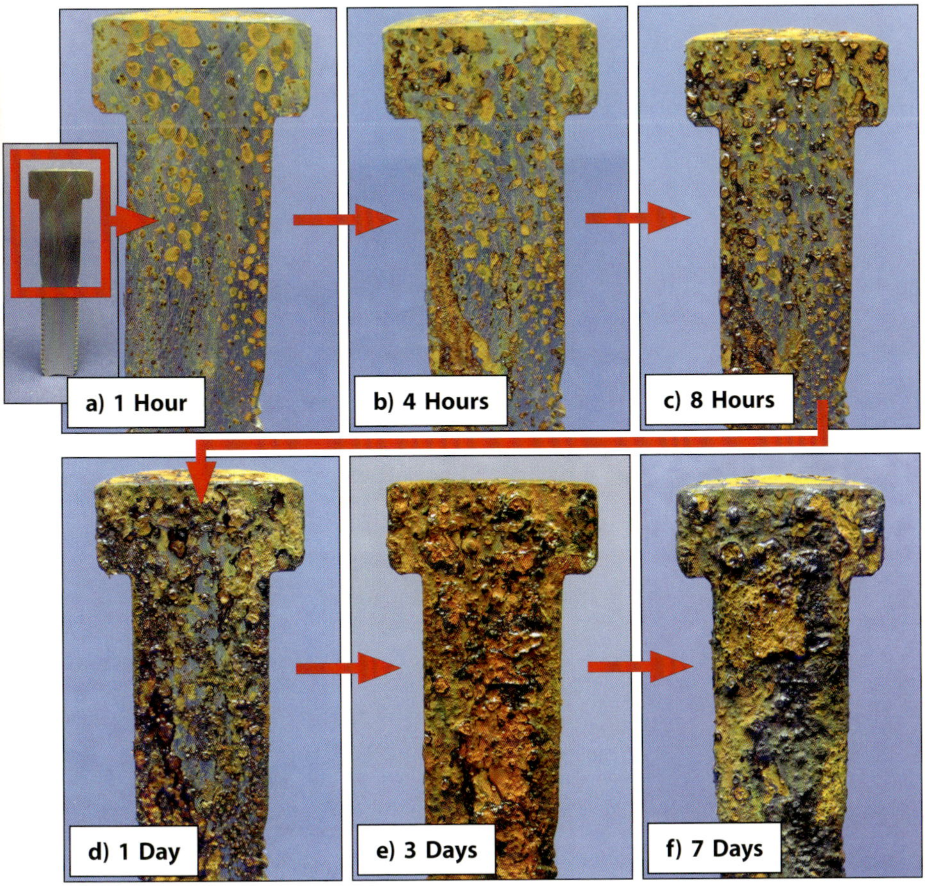

Figure 9.1: Water causes steel to corrode. These photographs display the progression of water corrosion during an accelerated corrosion test (salt water fog, ASTM B117). Corrosion is ugly, it consumes steel and the rust that forms expands. For steel products to remain functional and aesthetically pleasing, efforts must be made to prevent corrosion.

Water is the enemy of steel. It is responsible for the vast majority of steel corrosion. Once water has been in contact with air, it becomes corrosive to steel. **Figures 9.1** and **9.2** show how water consumes steel to form red-brown rust. Steel's susceptibility to water corrosion is the primary reason why constructions such as bridges and ships do not last forever. Therefore, it must be managed carefully.

Figure 9.2: Red-brown rust corrosion of a thirty-year old steel lamp post. Water had penetrated through the paint and had begun to corrode the steel. At this point, the corrosion had only been superficial. Yet with the paint now damaged, the rate of corrosion would increase and after another decade or two, might risk the integrity of the post.

Corrosion first affects the cosmetic appearance of steel – it just looks ugly. Few people will buy food cans or new cars that have already begun to rust. Yet the consequences of corrosion can be more troublesome. The conversion from steel to rust comes with a volume expansion – the rust that forms is many times the size of the steel that it consumes. For complex equipment, this rust expansion can cause moving components to seize or wear. With enough time and material loss, corrosion can cause parts and structures to fail.

In addition to simple water corrosion, impurities such as pollution, chemicals and salts can further accelerate the rate of corrosion. For example, the salt in sea water causes steel to corrode five times faster than fresh water. Stagnant conditions, erosion, crevices, bioactivity, electrical current or the joining of dissimilar metals can also affect the rate of damage. When water can directly touch steel, a lot can go wrong over time.

Table 9.1 lists the rates of steel corrosion in different environments. Note that the rates are not overly quick. To corrode through one inch of steel could take hundreds of years depending on its environment. Still, whether thin steel on car bodies or thick bridge beams, efforts must be taken to prevent or at least minimize corrosion. Otherwise, water damage could force retirement or risk failure.

Table 9.1: Typical Water Corrosion Rates of Steel[1]

Environment	**Thickness Loss Per Year**	**Years to Consume***	
		0.10 inches	**1 inch**
Dry, desert conditions	<0.001 inches	100+	1000+
Moderate climate, moderate rainfall	0.001 to 0.003 inches	33-100 years	330 to 1000 years
Industrial, polluted with sulfur	0.005 to 0.010 inches	10-20 years	100-200 years
Marine applications	~0.020 inches	5 years	50 years

[1] *API 571 – Damage Mechanisms Affecting Equipment in the Refining Industry.* American Petroleum Institute (2020).
* One side exposed.

General metal wastage from water exposure is responsible for the vast majority of steel corrosion damage. Yet exposure to chemicals, biological activity and hot temperatures can also cause more complex forms of damage – such as corrosion cracks, pitting or even damage internal to the steel. Some of these complex corrosion mechanisms occur much faster than simple water corrosion.

9.2 Preventing Water Corrosion

9.2.1 Coatings

Despite the harm that water can cause, steel is still successfully used for applications exposed to rain, coastlines, tropical conditions and even submerged within water. Think of all the cars, buildings, bridges, ships, submarines, oil rigs and pipelines in the world. Although there are several strategies that are used to prevent corrosion, the most common one is based upon a simple concept – do not let water touch steel. Provide a barrier.

There are a variety of coatings that are commonly used to create barriers that water cannot penetrate. As simple as it may seem, paint is by far the most common coating. If you look around, you will see many outdoor applications that use water-tight paint as the primary defense against corrosion. The second most common coating family are zinc-based coatings referred to as 'galvanized' coatings. These also provide water-tight barriers plus have the advantage of delaying steel corrosion by the zinc preferentially corroding first if the coating is breached (further discussed in *Section 9.2.4*).

Figure 9.3: Coatings create water barriers that enable the use of steel outdoors and even in marine environments. There are numerous types of coatings ranging from epoxy, zinc (galvanized), chrome and even other metals. The most common coating used can be seen almost everywhere you look – paint.

Make no mistake, water is a patient and relentless adversary. If there are flaws in the coating or it becomes damaged, water will find its way through and eventually attack the steel. Such as the stone chip damage on my car hood allowing for corrosion where the paint had been defeated (**Figure 9.4**). Ultimately, using coatings to prevent corrosion relies upon the quality, durability and maintenance of the coating applied.

Figure 9.4: Steel car bodies are protected by several coating layers including paint. Damage to the coatings can expose underlying steel and allow for the steel to rust. Stone chips such as these need to be repaired or else corrosion will eventually form holes through the thin steel. The coating's ability to protect steel depends upon its quality and durability in its environment of use.

9.2.2 Eliminating Dissolved Oxygen

When air touches water, small amounts of oxygen dissolves into the water. Small rain droplets falling through the sky become saturated with dissolved oxygen. The waves and the lapping of rivers, lakes and oceans mix oxygen into their waters. This dissolved oxygen is what fish breath through their gills. It is also what makes water corrosive to steel. Effectively all water that you can see or touch will contain dissolved oxygen and be harmful to steel.

However, water by itself without dissolved oxygen or impurities is not corrosive to steel. This means that if efforts are made to remove dissolved oxygen from water and prevent further air contact, the water will not harm steel. This concept of eliminating dissolved oxygen from the equation is possible for only a few enclosed applications. Boilers used at power plants or for heating buildings are filled with treated water from which free oxygen has been removed. Industrial buildings' sprinkler systems, which often remain filled with stagnant water for years or decades, will stop corroding once all the dissolved oxygen in the water is consumed by rust.

9.2.3 Chemicals

For short-term exposure, chemicals such as rust inhibitors can be added to water. For example, the water-based fluids used during the cutting or machining of steel parts include rust inhibitors. This prevents corrosion during manufacture.

9.2.4 Cathodic or Anodic Protection

If a small yet precise current is applied to steel, it can prevent steel corrosion. This can be completed by either (a) applying a carefully designed 'cathodic' electrical current or (b) connecting 'anodic' zinc metal to the steel. The latter, anodic option protects steel by causing the zinc to corrode instead. This anodic benefit is why zinc-based (galvanized) coatings are popular – they provide both water-tight barriers and if damaged, will delay steel corrosion by the zinc corroding first.

9.2.5 Switching Materials

There are other metals that are more corrosion resistant than conventional steel. These include metals such copper, brass, stainless steel, nickel and aluminum. If the corrosion prevention strategies available to steel do not provide enough protection, or the environment is just too harsh, changing to a more corrosion resistant material may be necessary.

9.3 Summary of Water Corrosion

1. Water is the relentless enemy of steel – combined with even small amounts of dissolved oxygen, it causes steel to corrode.

2. Corrosion is ugly, corrosion products expand and corrosion slowly consumes steel.

3. Although water-based corrosion is the most common form of corrosion, exposure to chemicals, electrical current, bioactivity and high temperatures can cause faster rates of corrosion and/or more complex damage.

4. Common corrosion prevention strategies:

    a. Prevent water from touching steel – apply coatings such as paint.

    b. Limit oxygen – clean water without dissolved oxygen (from air exposure) will not corrode steel.

    c. Chemicals – rust inhibitors can be applied to water.

    d. Cathodic or anodic protection – applying current or connecting zinc to steel.

    e. Switch to a more corrosion resistant material – stainless steel, copper, brass, etc.

With over 11% chromium, steel becomes 'stainless' and can resist water corrosion.

10.0 STAINLESS STEEL

10.1 Enough Chromium Makes It Stainless

The red-brown rust that steel forms from water corrosion is porous, allowing water to seep through and constantly attack the steel beneath. Yet if the chromium content within steel is increased to above 11%, it becomes 'stainless' and resistant to corrosion. At this chromium threshold, the steel forms a protective layer that blocks corrosion from occurring. **Figure 10.1** illustrates how a stainless steel bolt with over 11% chromium was essentially immune to the water-based test conditions that caused a carbon steel bolt to rust.

Figure 10.1: Unlike conventional steels that quickly corrode during accelerated salt-fog corrosion testing (ASTM B117), the high chromium of stainless steels make them resistant to corrosion in this environment.

As always, the properties of steel are based upon the same two principles: composition and microstructure. To be considered stainless steel and self-protecting against corrosion, it must have at least 11% chromium. To further increase the corrosion resistance or to modify other properties, additional chromium and other alloying elements are often included. **Table 10.1** lists five classes of stainless steels. Each are named after their prevalent microstructure – another indication of how important microstructure is. For this introduction, we will discuss the three most common classes in more detail.

Table 10.1: Common Classes of Stainless Steel

Class	Typical Alloying	Common SAE Grades	Description
Austenitic	17-20% chromium 7-10% nickel 0-3% molybdenum	304L, 316L	• Most common stainless steel. • Good corrosion resistance, easy to weld/ manufacture. • Moderate strength.
Martensitic	11-15% chromium 0.15-0.40% carbon	410, 440	• High strength. • Moderate corrosion resistance.
Ferritic	11-12% chromium	409	• Cheapest. • Poorest corrosion resistance of the stainless steels. • Weakest.
Duplex (mixed austenite and ferrite)	21-25% chromium 5% nickel 2-5% molybdenum	2205	• Higher corrosion resistance than austenitic stainless steel. • More durable in chlorine-containing environments.
Precipitation Hardened	15-17% chromium 3-5% copper or aluminum	17-4, 15-5	• High strength and corrosion resistance.

10.2 Austenitic Stainless Steels (Most Common)

The most common stainless steel is '304L' which has 18% chromium and 8% nickel. It has good corrosion resistance, moderate strength and is easy to use during fabrication. Its high chromium content prevents corrosion in many water environments and makes it safe to handle moderately aggressive chemicals.

Chromium provides its corrosion protection while nickel is added for another reason. The 8% nickel is enough to form a microstructure of 'austenite' (**Figure 10.2**). This phase is different from the ferrite, pearlite and martensite that form in carbon and low alloy steels. The austenite microstructure makes the metal ductile and easy to form into shape. Combined with its low carbon, 304L is also easy to weld (the 'L' designates it as 'low carbon').

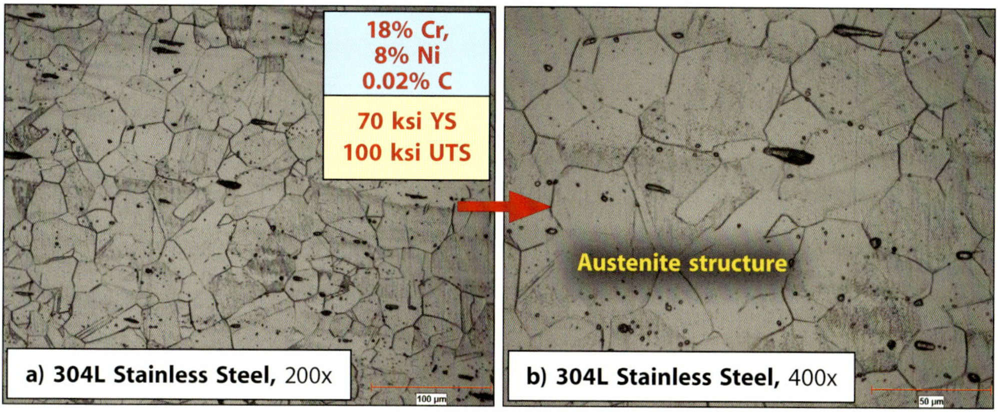

Figure 10.2: Due to its 8% nickel, 304L stainless steel does not form the same microstructural phases as seen in conventional carbon and low alloy steels. Instead, it forms an austenite microstructure which is easy to cut, shape and weld. Note that austenite looks similar to ferrite under a microscope yet requires different chemicals to etch. Electrolytically etched using 10% oxalic acid.

If 304L does not provide enough corrosion resistance, 316L stainless steel is even more resistant to corrosion. In addition to similar chromium and nickel contents, 316L also contains 2% molybdenum which further improves its corrosion resistance. This makes it suitable for a wider range of chemicals, seawater and other applications that include more harmful substances.

Austenitic stainless steels are used for applications that require good corrosion resistance and, in order to make the parts, require ample ductility to shape and must be easy to weld. This covers a wide range of applications spanning from food processing, chemical handling, marine applications to even the body of the SpaceX Starship (**Figure 10.3**).

Figure 10.3: Their corrosion resistance and ease to manufacture with makes austenitic stainless steels ideal for many applications including food processing, chemical handling and even space projects.

Note that no matter how robust a class of stainless steel may seem, each grade has a weakness. For instance, chlorine dissolved in water is the kryptonite of austenitic stainless steel. Heated water with chlorine/chlorides can cause pitting and even corrosion cracking of austenitic stainless steel.

10.3 Martensitic Stainless Steels (Strongest)

This class of stainless steel provides the combination of moderate corrosion resistance with high strength. They often have 11% to 15% chromium with at least 0.15% carbon. Their combined chromium and carbon contents make them hardenable, forming high strength martensite when quenched and tempered. **Figure 10.4** displays the structure formed within a quenched and tempered 410 martensitic stainless steel. Its martensite behaves in a similar manner as hardened carbon and low alloy steels – including how the tempering temperature will adjust the carbide distribution and strength properties (**Figure 10.5**). These steels are difficult to weld and, once hardened, difficult to machine or shape.

Martensitic stainless steels are used for applications that require either high strength or wear resistance (resistance to dulling) in combination with corrosion resistance. Common personal items include kitchen knives, razor blades and ice skate blades (**Figure 10.6**).

Figure 10.4: Martensitic stainless steels can be quenched and tempered into martensite, experiencing similar strength benefits as hardenable carbon and low alloy steels. This 410 stainless steel had been water quenched and then tempered at 1000°F (538°C). Etched using Vilella's Reagent.

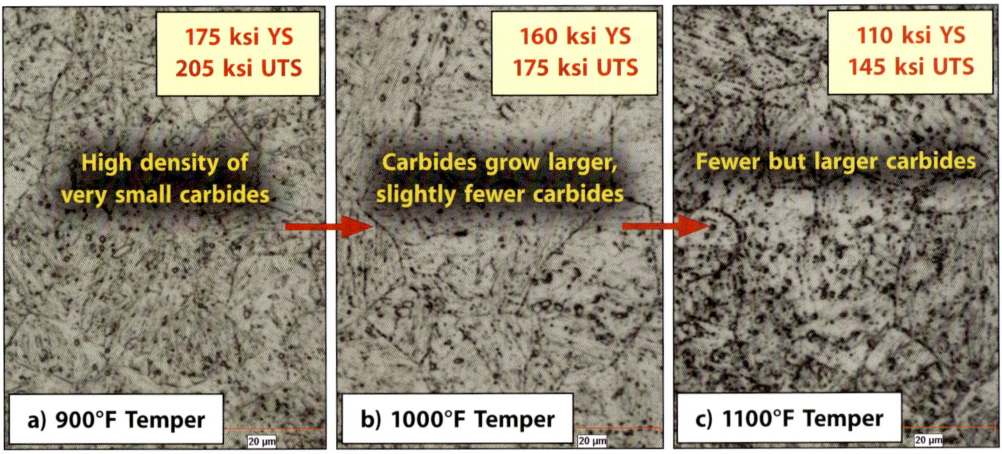

Figure 10.5: Similar to carbon and low alloy steels, the tempering temperature affects the carbide size and distribution within martensitic stainless steels, adjusting their final properties. 410 stainless steel, etched using Vilella's Reagent.

Figure 10.6: Martensitic stainless steels are useful for applications that need to resist corrosion and require either high strength or resistance to dulling. Knives, skate blades and razor blades are common martensitic stainless steel items that you may own.

10.4 Ferritic Stainless Steels (Cheapest)

This is the cheapest class of stainless steel, giving you only what you pay for – the bottom threshold of corrosion resistance available to stainless steels. It contains the bare minimum 11% chromium content needed to be considered stainless. Its microstructure comprises solely of ferrite which, like the ferrite in conventional steels, provides relatively low strength (**Figure 10.7**).

Its ample ductility and low carbon make ferritic stainless steel easy to form and weld. Combined with the bottom threshold of corrosion resistance available to stainless steel, this family is used for applications such as car mufflers and indoor appliances (**Figure 10.8**).

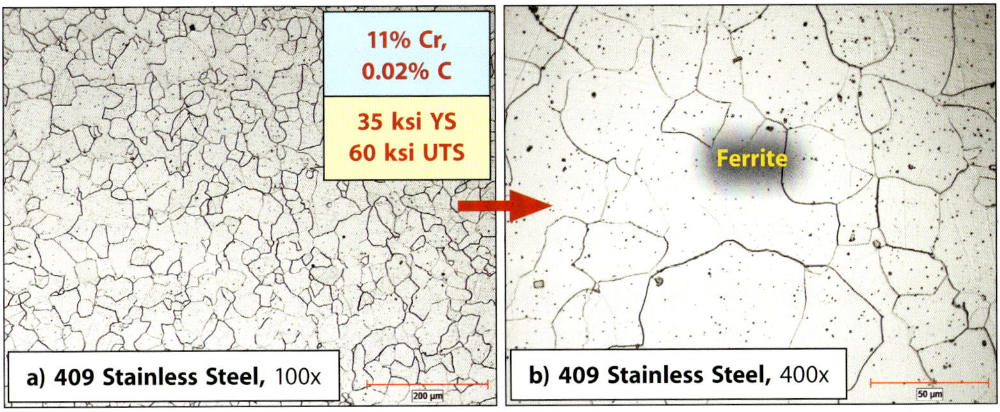

Figure 10.7: Ferritic stainless steel has just enough chromium to be called stainless. Its low carbon content and ferrite structure leaves it rather weak. Yet it is easy to shape and weld. 409 stainless steel, etched using Vilella's Reagent.

Figure 10.8: The combination of relatively low cost, ease to manufacture with and its corrosion resistance (low for stainless) makes ferritic stainless steels useful for indoor appliances and out-of-sight mufflers.

10.5 Summary of Stainless Steel

1. Steels with chromium contents greater than 11% form a self-protecting surface layer that prevents corrosion. This makes the steel 'stainless'.

2. Different classes of stainless steels are designed to optimize different properties. Each class is named after its predominant microstructure.

3. The three most common classes of stainless steel are:

    a. Austenitic stainless steel – good corrosion resistance, moderate strength, easy to weld and manufacture with.

    b. Martensitic stainless steel – highest strength, moderate corrosion resistance, difficult to weld and, once hardened, difficult to machine.

    c. Ferritic stainless steel – cheapest, bottom threshold of corrosion resistance for stainless steel, easy to form and weld.

Understanding steel has allowed humankind to build the world we live in today.

11.0 SUMMARY

The better humanity has understood steel, the more incredible our creations have become. It has allowed us to construct the buildings and bridges that shape our cities. We have used steel to produce the cars, trains and ships that connect our communities. Even within our homes, our daily habits are fashioned by the steel appliances, cookware and tools that we keep. Our modern way of life is built upon the intelligent use of steel.

The reason why steel is so useful is because of its relatively low cost, high strength and its ability to be tailored to suit a wide range of applications. This latter trait is accomplished by adjusting the steel alloying and manufacturing steps to optimize its properties for a particular use. This is why the ability to design, manufacture and even to repair steel components requires a practical understanding of its behavior.

The properties of steel are based upon its alloying and microstructure. The most critical alloying element of steel is carbon. Adding carbon increases its strength and, if enough is present, small parts can be quenched to even higher strengths. Molybdenum, chromium and nickel further strengthen the steel and increase the part size that can be strengthened by quenching.

The drawback of high carbon content is that it makes steel difficult to manufacture with. To mitigate this, higher carbon and well alloyed steels are often machined and shaped before being heat treated to their final strengths. Yet steels with ample carbon and alloying will always be more burdensome to weld than steels with low carbon.

Once the steel grade is selected and the alloying set, the rest is up to its microstructure. Most professionals will never look down a microscope and see a steel's microstructure for themselves. Yet being aware of the role that microstructure plays is the only way to understand how heat treating, processing and welding affect the properties of steel.

Different microstructures produce different properties. The common microstructures ranked from weakest to strongest are ferrite and pearlite, bainite, then martensite. The microstructure within a particular steel is based upon its composition and its processing – starting with whether it had been slowly cooled or quickly quenched from red-hot temperatures during heat treatment.

Table 11.1 summarizes the effects that alloying and cooling rates had upon the three bolt grades referenced throughout this book. Air cooling during normalizing formed ferrite and pearlite microstructures, providing moderate strength. For the bolts with over 0.3% carbon, water quenching had formed high strength martensite. As-quenched martensite is brittle and unsafe. Yet tempering returned its ductility and adjusted the strength properties – a useful trait for customizing steel for a particular application.

There are many methods to shape and transform raw steel into functional components and structures. For each of these, steel behavior needs to be considered so the finished part or construction has both the intended geometry and properties. This includes welding which, depending upon the steel grades being joined, may require additional steps to produce a robust weld.

As amazing as steel is, be aware of its vulnerability to its arch nemesis: water. The corrosion that water causes is ugly, the rust formed can bind moving components and it can eat away at the steel thickness. Different strategies exist to prevent corrosion, the most common being to provide a barrier and prevent water from touching steel. Yet if coatings such as paint are not sufficient, changing to a more corrosion resistant material may be required.

With a minimum chromium content of 11%, steel becomes ‘stainless’ and self-protecting against corrosion. The different classes of stainless steels are designed to have different benefits; some are easier to manufacture and weld while others are optimized for strength or cost.

Steel can be quite strong and hard – these are amongst the reasons why we use it. Yet hopefully you can now appreciate that steel is not hard to learn. This book has provided an introduction to steel by explaining the most critical and most pragmatic concepts relevant to its use. Many of the steel components you will see and touch can be understood by these fundamentals. However, consider this as only the foundation to build your expertise upon. Each role in each industry will use different materials, construction methods and face different challenges – each requiring specific proficiencies and know-how. Yet whether designing, manufacturing or repairing, the more you learn about steel and the other materials you work with, the better you will be at what you do.

Table 11.1: Effects of Alloying and Cooling Rates on Microstructure and Strength

Bolt Grade	Notes	**Annealed** 1650°F then slowly furnace cooled
Low Carbon 0.15% C (Grade 2 Bolt)	\$0.45 per bolt Not hardenable Easy to weld	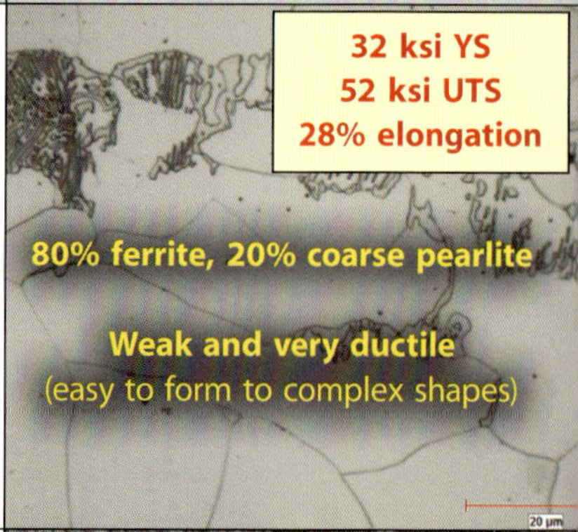 32 ksi YS 52 ksi UTS 28% elongation 80% ferrite, 20% coarse pearlite Weak and very ductile (easy to form to complex shapes)
↓ Increase in carbon		
Medium Carbon 0.4% C (Grade 5 Bolt)	\$0.60 per bolt Thin parts hardenable Time consuming to weld	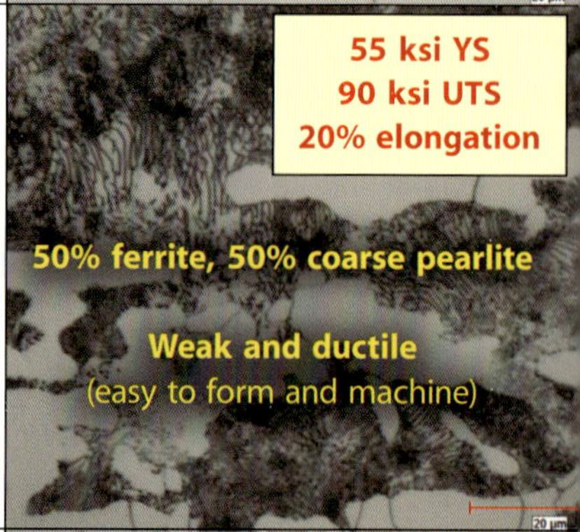 55 ksi YS 90 ksi UTS 20% elongation 50% ferrite, 50% coarse pearlite Weak and ductile (easy to form and machine)
↓ Addition of Mo, Cr, Ni		
Low Alloy 0.4% C +Mo+Cr+Ni (Grade 9 Bolt)	\$1.50 per bolt Thick parts hardenable Time consuming to weld	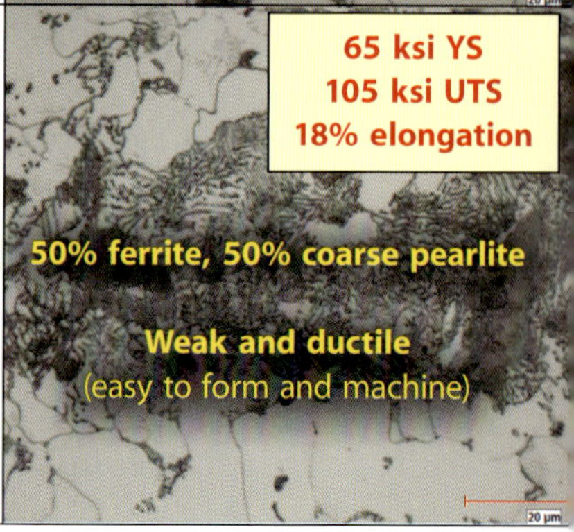 65 ksi YS 105 ksi UTS 18% elongation 50% ferrite, 50% coarse pearlite Weak and ductile (easy to form and machine)

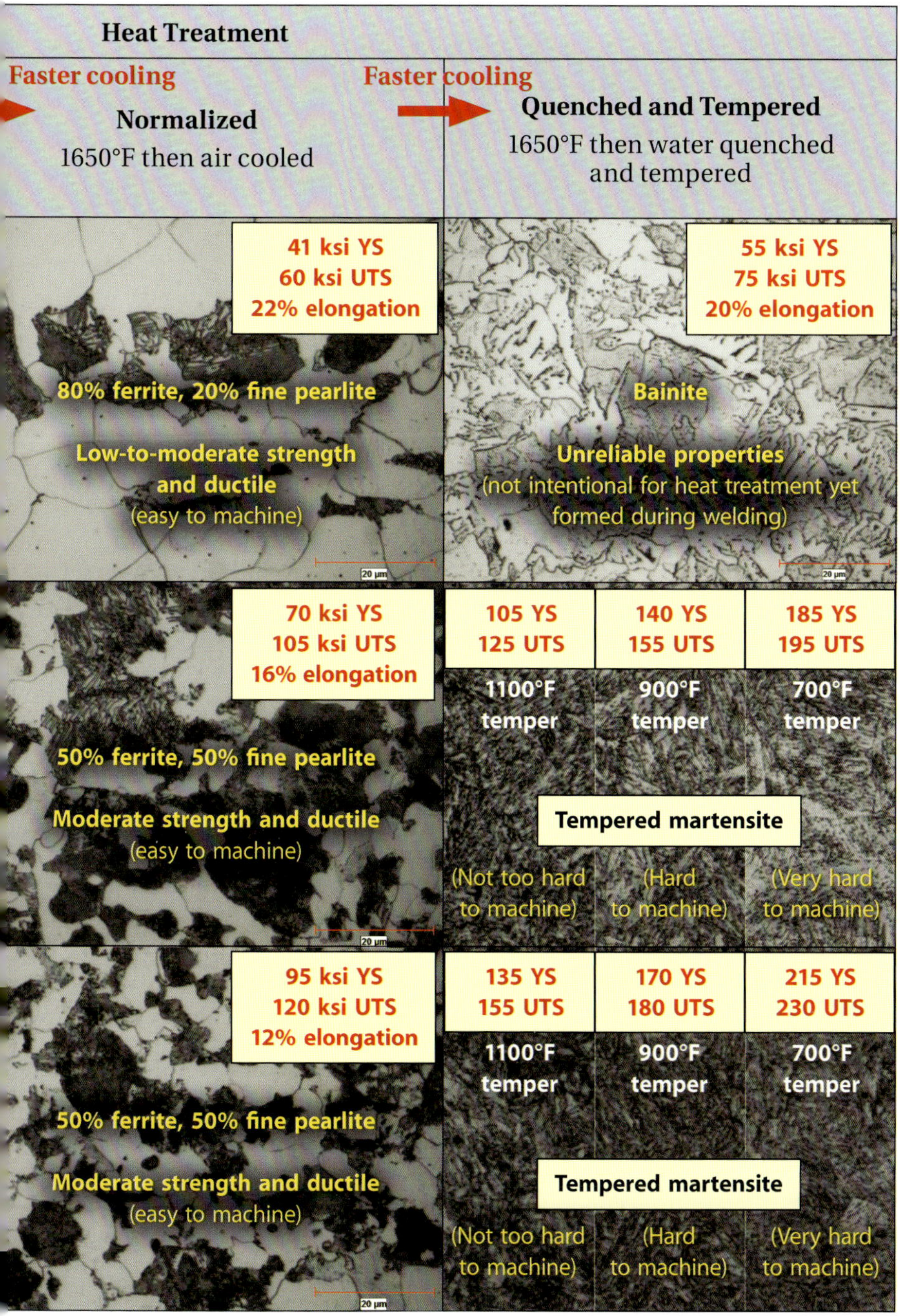
Heat Treatment
Faster cooling
Normalized
1650°F then air cooled
Faster cooling
Quenched and Tempered
1650°F then water quenched and tempered
41 ksi YS
60 ksi UTS
22% elongation
80% ferrite, 20% fine pearlite
Low-to-moderate strength and ductile
(easy to machine)
20 µm
55 ksi YS
75 ksi UTS
20% elongation
Bainite
Unreliable properties
(not intentional for heat treatment yet formed during welding)
20 µm
70 ksi YS
105 ksi UTS
16% elongation
50% ferrite, 50% fine pearlite
Moderate strength and ductile
(easy to machine)
20 µm
105 YS
125 UTS
1100°F temper
(Not too hard to machine)
140 YS
155 UTS
900°F temper
(Hard to machine)
185 YS
195 UTS
700°F temper
(Very hard to machine)
Tempered martensite
95 ksi YS
120 ksi UTS
12% elongation
50% ferrite, 50% fine pearlite
Moderate strength and ductile
(easy to machine)
20 µm
135 YS
155 UTS
1100°F temper
(Not too hard to machine)
170 YS
180 UTS
900°F temper
(Hard to machine)
215 YS
230 UTS
700°F temper
(Very hard to machine)
Tempered martensite

APPENDICES

These appendices include optional, more technical discussion touching upon the science behind steel behavior. Much of this content goes beyond the practical aspects needed for working with steel.

Studying how steel behaves during tensile testing can be used to describe how steel responds to stress.

APPENDIX A: STRESS-STRAIN CURVES AND BEHAVIOR UNDER STRESS

Each stress-strain curve produced by tensile testing is a precision recording of how that particular steel sample responded to being pulled apart. Not only do these diagrams provide key strength values, they describe how steel behaves under stress or strain. The following is a detailed, technical discussion for those interested in better understanding how steel strength is measured and how steel reacts under stress.

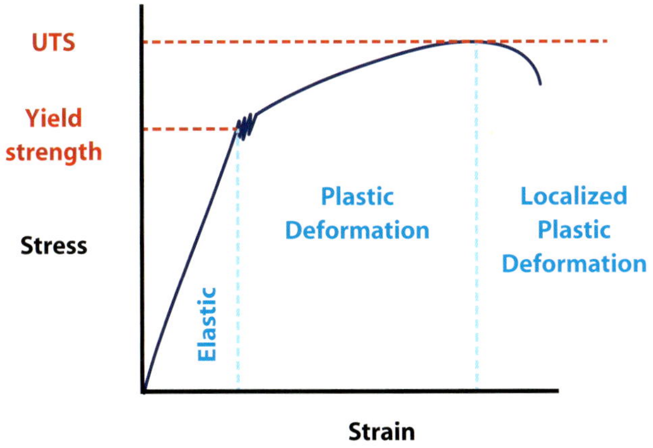

Figure A.1: Stress-strain curves not only provide critical strength values, they are also recordings of how steel reacts while being pulled apart. They can be used to explain how steel behaves under both stress and strain.

A.1 Stress and Strain – Measuring, Calculating and Plotting

As mentioned in Chapter 3, a tensile testing machine slowly pulls apart a machined specimen while measuring the resistance it provides. The stronger the steel, the greater the force needed to pull it apart. The stretching of the tensile specimen, reported as 'strain', is first measured by an extensometer and then by crosshead displacement (**Figure A.2**). The force of resistance that the steel specimen provides is measured and converted into 'stress' by dividing out the load bearing area.

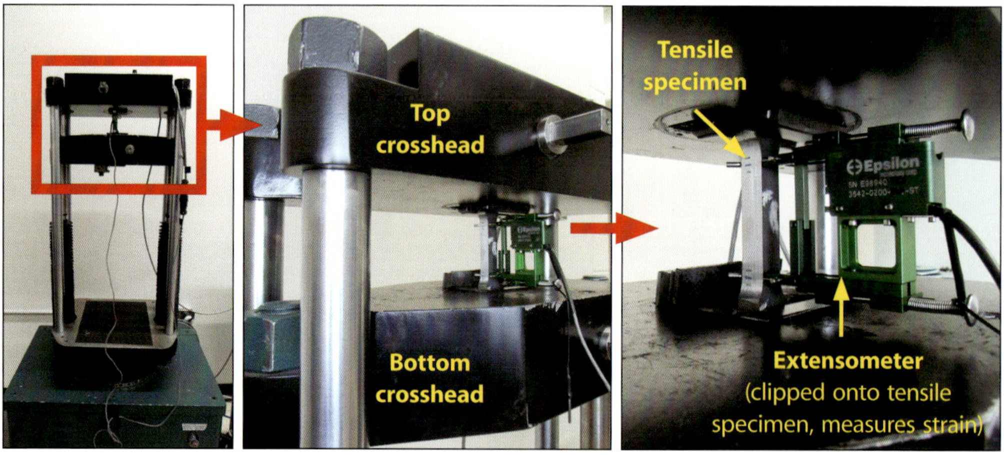

Figure A.2: During tensile testing, strain is first measured by an extensometer. There are several types; the most common extensometer design clips onto the tensile specimen and, as it stretches, the change in clip spacing is measured. Once past the yield strength, the same degree of accuracy is no longer needed. At that point, the extensometer is removed and the remainder of the strain is calculated from crosshead movement.

If you feel comfortable with equations, stress and strain can be described as listed in **Table A.1**. Yet grasping the math is not required to understand the significance of each.

Once the data from tensile testing is converted into stress and strain, they are plotted to create 'stress-strain curves' as shown in **Figure A.1**.

Table A.1: Stress and Strain Definitions

Stress (σ)	**Strain (ε)**
Load per unit area	Ratio of stretch
$Stress = \dfrac{Load}{Load\ Bearing\ Area}$	$Strain = \dfrac{Increase\ in\ length\ (\Delta L)}{Original\ length\ (L_i)}$

Although tensile tests apply strain and measure stress, the curves they produce work both ways – they describe how the steel responds to either applied strain or stress. Therefore, we can use stress-strain curves to break down and describe the stages that steel goes through under load.

A.2 Elastic Behavior

During the initial stage of loading, at stress levels below the yield strength, the steel behaves in an 'elastic' manner (**Figure A.3**). At these stress levels, if the stress or strain were removed, the steel would spring back to its original shape. Within the elastic regime, the atoms do not move relative to one another – only the bonds between the atoms stretch to accommodate. Upon unloading, the atomic bonds spring back and everything returns to its original dimensions. Since there is no atomic movement, there is no change to the microstructure.

Almost all steel parts are designed to operate below their yield strength and within the elastic range. Even flexible parts like suspensions or springs that bend under load, provided they stay within the elastic range as intended, will return to their original shape once unloaded.

In the elastic region, the relationship between stress and strain is linear – forming a straight line on the stress-strain curve. The correlation between how much the steel strains under a given stress is described by the 'Elastic Modulus' which can be calculated as shown in **Figure A.3**.

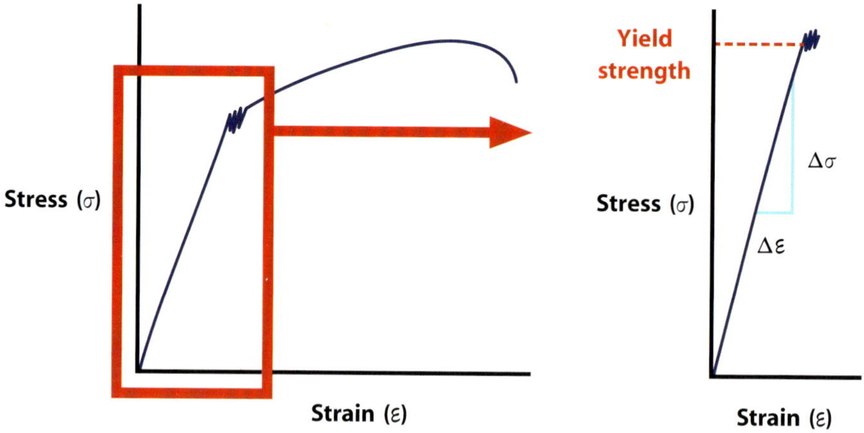

$$Elastic\ Modulus\ (E) = \frac{\Delta\sigma}{\Delta\varepsilon}$$

Figure A.3: Below the yield strength, the steel behaves in an elastic manner. When stressed, it strains in a linear manner (or vice versa). The ratio of stress-to-strain is based upon the steel's 'Elastic Modulus'.

The Elastic Modulus for all metals is based solely on its composition. Since conventional steels comprise of over 98% iron, most steels have essentially the same Elastic Modulus of 29,000 ksi. This means that the steel strength, microstructure and heat treatment does not affect the modulus; the ratio of stretch or deflection for all conventional steels will be the same within the elastic regime. For example, any carbon or low alloy steel experiencing 30 ksi of stress will stretch just over 0.1% as calculated below. Steel would stretch twice as much if the stress were doubled provided it was still within the elastic regime.

$$Strain = \frac{Stress}{Elastic\ Modulus\ (E)} = \frac{30\ ksi}{29{,}000\ ksi} = 0.00103\ or\ \sim 0.1\%$$

Due to its high Elastic Modulus, steel is very rigid and incredible amounts of stress are required to cause tiny amounts of stretch. Even under 30,000 lbs of load, the stretch in a hardened 5/8-inch Grade 9 bolt would not be visible by eye. Yet for longer parts, such as long bridge beams, elastic deflection may be felt during severe storms or earthquakes.

A.3 Yielding

The elastic regime ends when steel sustains enough stress or strain that the atoms begin to move and slide relative to one another. Described as 'yielding', this alters how the steel responds to further increases in stress or strain – it will no longer be linear.

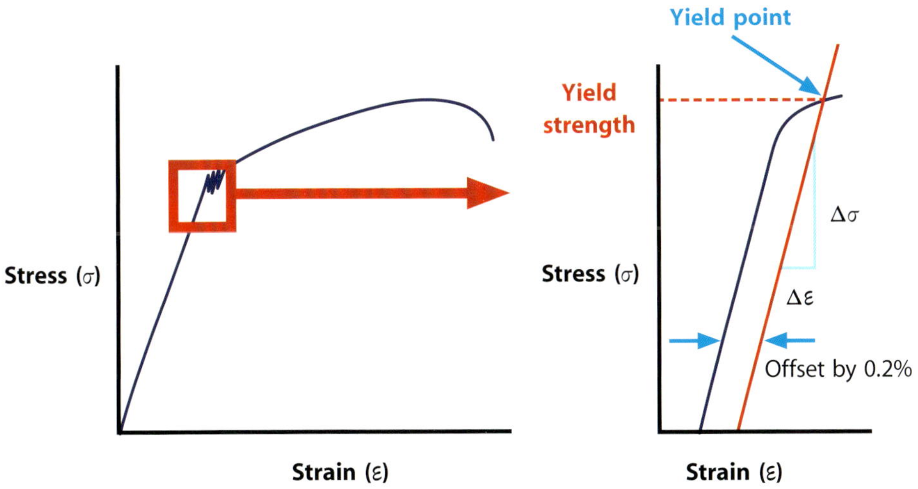

Figure A.4: The 'yield point' is determined when the plot deviates from the straight, linear behavior described by the Elastic Modulus and intersects with a parallel line offset by 0.2% strain. The stress at the point of intersection is the 'yield strength'.

The onset of yielding is detected when the stress-to-strain ratio begins to deviate from linear behavior. This is accomplished by drawing a parallel line offset by 0.2% from the elastic stress-strain line (**Figure A.4**). When the steel begins to yield and its behavior changes, it intersects this offset line. That intersection marks the 'yield point' and the corresponding stress is considered the 'yield strength'.

At the point of yielding, the atoms and microstructure will move to accommodate any additional stress or strain. This means that further stress or strain will induce permanent shape change and the part will never again spring back to its original shape.

The transition from elastic behavior to beginning to deform is not always smooth. The initial movement of the atoms may be jerky and intermittent, creating chatter in the stress-strain curve as illustrated in **Figure A.5**. This unstable transition is referred to as 'yield point elongation'.

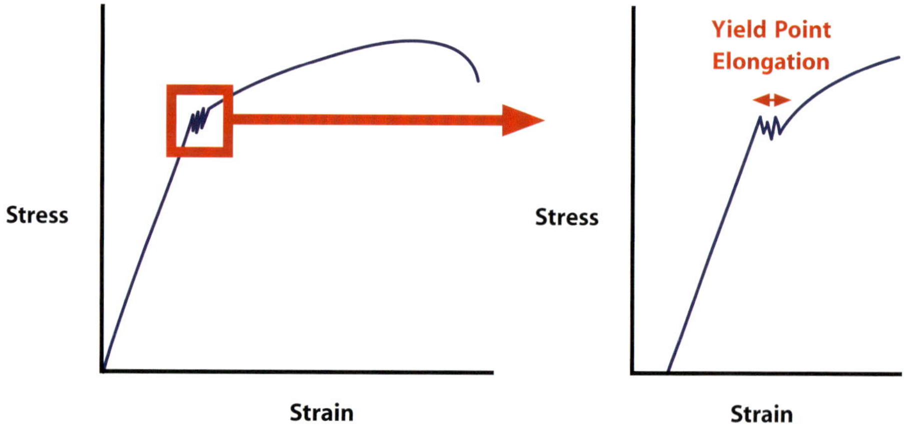

Figure A.5: The transition from elastic behavior can include chatter referred to as 'yield point elongation'. Due to clusters of atoms beginning to move at marginally different stress thresholds, the onset of atomic movement is sometimes intermittent and can create a choppy transition to deformation behavior.

Yield point elongation is of little concern for finished parts since the service loading should never reach their yield strengths. If the loading on a part does accidentally exceed its yield strength, no one cares whether it was a bumpy transition or not as deformation causes the part to cease functioning. However, it can sometimes be a problem during manufacture when cold forming parts to shape. If this is a concern during forming/manufacture, there are methods of pre-straining the steel to eliminate yield point elongation.

A.4 Plastic Deformation

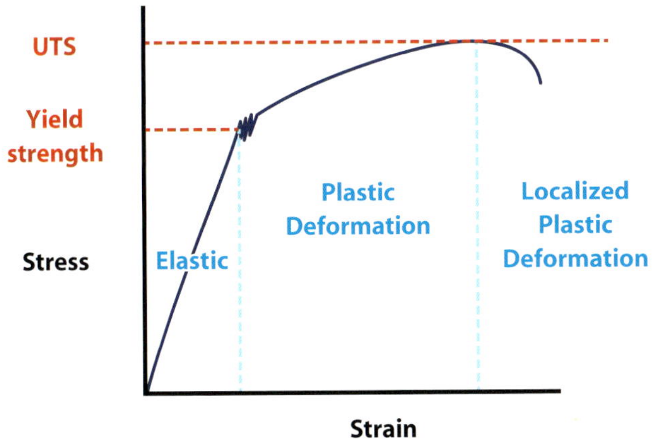

Figure A.6: At stresses or strains beyond the yield point, the steel permanently stretches. The atoms and microstructure shift to accommodate increasing stress and strain.

The permanent shape change that occurs after yielding is referred to as 'plastic deformation'. **Figure A.7** illustrates a tensile specimen that had been pulled beyond its yield point and well into the plastic regime. When unloaded, it was clear it had been permanently stretched. Although useful during manufacturing and the cold forming of parts, few finished parts are intended to deform during use. Examples of unwanted deformation include everything from dents in a car door, bending of a spoon or worse, buckling and collapsing of a crane.

If a steel tensile specimen were unloaded after being plastically deformed, there would be a tiny amount of contraction following along the Elastic Modulus slope of 29,000 ksi. **Figure A.8** illustrates this elastic springback. **Figure A.8** also shows that, if reloaded again, the yield strength would have increased to the former stress level it had been previously deformed to.

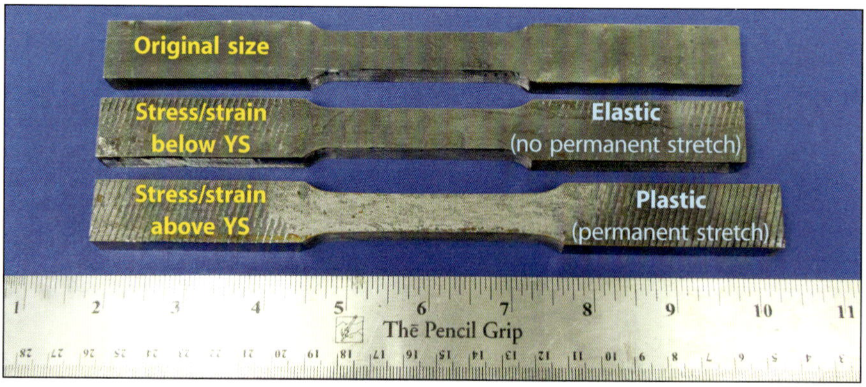

Figure A.7: When stressed or strained beyond its yield point, steel begins to experience permanent deformation. In the case of a tensile specimen, the narrower center length begins to stretch. Since this stretching deformation is irreversible, it is referred to as 'permanent deformation' or 'plastic deformation'.

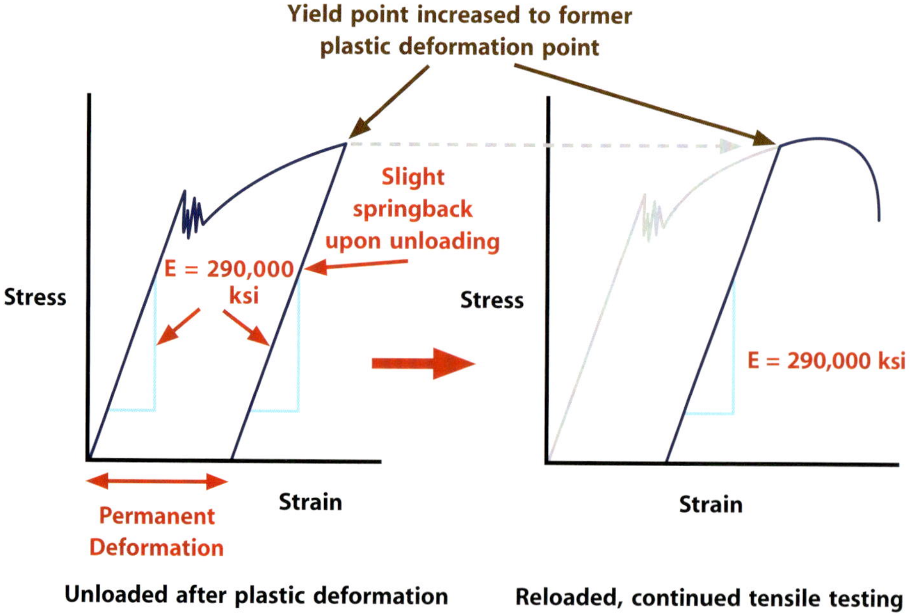

Figure A.8: (Left) If a tensile specimen were unloaded after sustaining plastic deformation, a small degree of springback would occur, corresponding to its Elastic Modulus. (Right) If that tensile specimen were then reloaded, its yield strength would have increased to the stress from which it had been unloaded. This yield strength increase comes at the loss of remaining ductility.

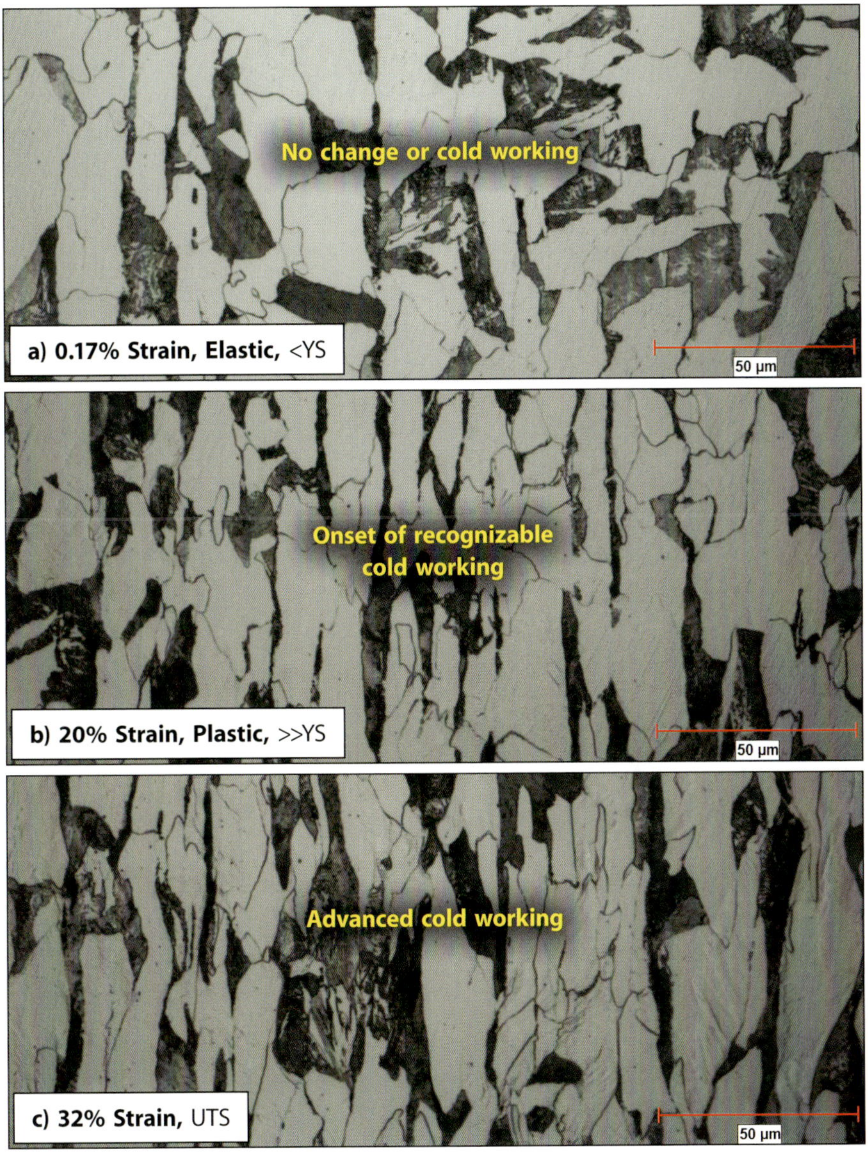

Figure A.9: (a) At stress levels below the yield strength and within the elastic regime, no change to the microstructure occurs. (b,c) Above the yield strength and in the plastic regime, the microstructure 'cold works' by elongating to accommodate the stretching deformation. Although its effects are immediate upon exceeding the yield strength, cold working becomes resolvable under a microscope at about 20% deformation. It is this cold working that increases the yield strength. Images of tensile tested ASTM A36 carbon steel (0.15% C).

The reason why the yield strength increases is because of what the plastic deformation does to the steel. The atomic movement associated with plastic deformation 'cold works' the microstructure (**Figure A.9**). It is this cold working from atomic shifting that remembers previous deformation and increases the yield strength.

Obviously steel can only stretch so far. At some point it will hit the maximum strain it can sustain. This corresponds to the peak of the stress-strain curve and represents the ultimate tensile strength (UTS). This stress threshold will cause that particular steel to rip apart. If steel is stressed to its UTS in service, it will immediately break.

As the steel breaks apart, its continued deformation becomes focused within a small portion of the tensile specimen (**Figure A.10**). A single site along the length of the tensile specimen will begin to contract and stretch. Referred to as 'necking', this is the final death throe before the steel specimen breaks apart.

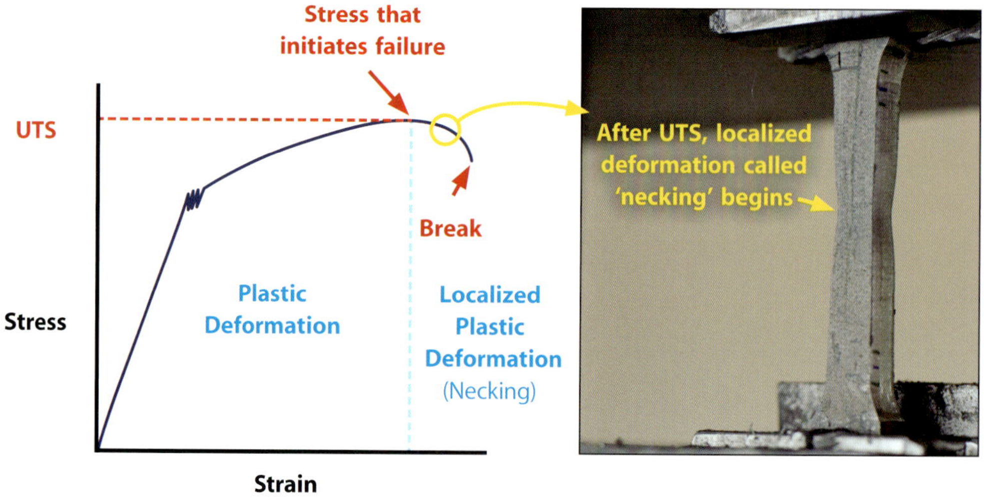

Figure A.10: At some point within the plastic regime, the tensile specimen begins to initiate failure and the resistance it provides begins to decrease. The point of this inflection is the UTS and corresponds to the stress threshold that would cause immediate failure. In the process of failing, the plastic deformation becomes localized. This localized stretching and reduction in cross-section is referred to as 'necking'.

Figure A.11 shows tensile specimens that had been stopped at different stages during tensile testing, plus one that had been pulled to complete failure. They summarize how steel responds to applied strain throughout tensile testing.

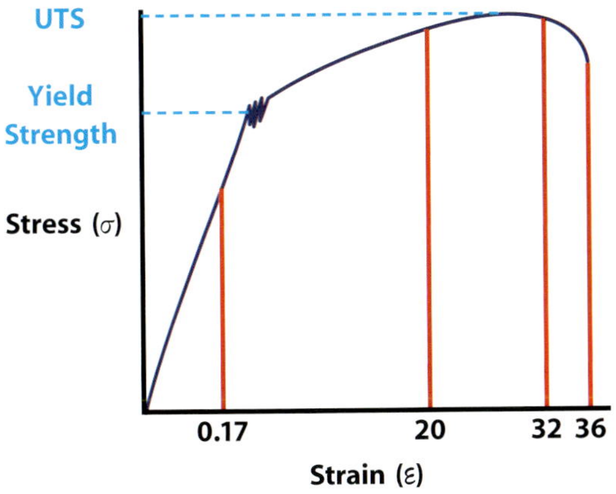

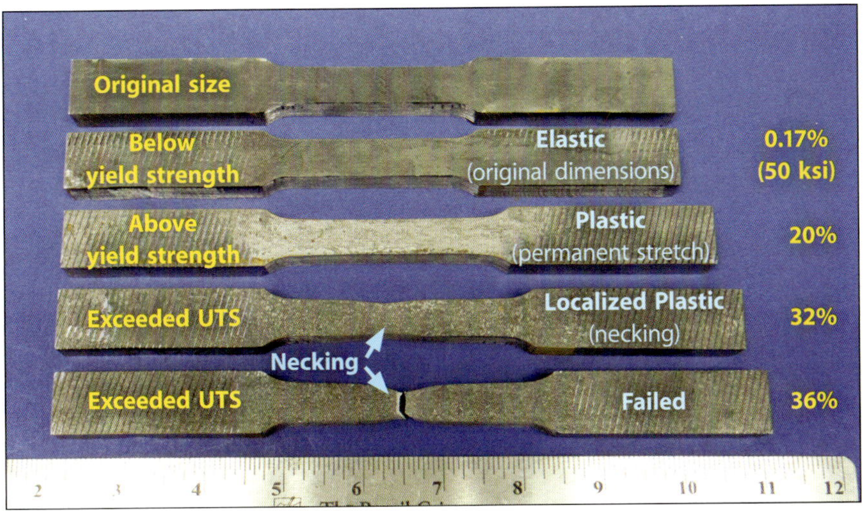

Figure A.11: Low carbon steel (0.15% C, ASTM A36) tensile specimens that had been stopped after different degrees of strain had been applied. Since 0.17% strain (12,500 lbs or 50 ksi) was still below its yield strength, that tensile bar had remained in the elastic region and no permanent stretch had been introduced. After 20% strain, the steel was far past the yield point and had been permanently deformed. At 32% strain, the steel had exceeded its UTS and had begun to neck. Final failure had occurred at 36% strain.

A.5 Summary of Stress-Strain Curves and Behavior Under Load

1. Stress-strain curves obtained during tensile testing:

    a. Are used to determine the yield and ultimate tensile strengths of steel.

    b. Describe how steel behaves under stress and strain.

2. At stress levels ranging up to its yield strength, steel behaves in an 'elastic' manner:

    a. Steel will spring back to its original shape when unloaded.

    b. The tiny amount of stretch that occurs is predictable, linear and described by its Elastic Modulus.

    c. Steel is quite rigid due to its high Elastic Modulus. Most conventional steels have an Elastic Modulus of 29,000 ksi.

    d. Most steel components are designed to operate within the elastic regime.

3. At stress levels above its yield strength, steel behaves in a 'plastic' manner and will become permanently deformed.

4. The ultimate tensile strength (UTS) is the stress threshold that will cause immediate failure.

    a. In the process of failing, a portion of the steel will begin to locally deform and neck as it breaks apart.

Each class of steel uses different alloying and microstructural strategies to either optimize one property or balance several.

APPENDIX B: COMMON STEEL GRADES – ALLOYING AND MICROSTRUCTURE

The generic effects of alloying and microstructure for carbon and low alloy steels have been discussed throughout this book. Comprehending the behavior of these two classes sets the foundation for understanding most types of steel. This section briefly introduces common grades of carbon and low alloy steel plus several extra classes of steel that you might see during your career.

B.1 Carbon Steel and Low Alloy Steel

Carbon steel is by far the most produced steel class in the world. Carbon provides the majority of its strength while manganese is a distant secondary strengthener. Carbon steels are divided into subcategories of low carbon (<0.3% C), medium carbon (0.3-0.6% C) and high carbon (0.6-0.8% C). Efforts are made during steelmaking to minimize the impurities of sulfur and phosphorus. **Table B.1** lists the compositions of the most common carbon steel grades. As already mentioned in Chapter 6, the initial '10' of the four-digit SAE designation allocates it as carbon steel while the last two digits mark the target carbon content.

Table B.1: Common Carbon Steel Grades

SAE Grade	Nominal Composition* (%)			
	C	Mn	S	P
1005	0.05	0.30	0.030 max	0.050 max
1010	0.10	0.45		
1015	0.15	0.45		
1020	0.20	0.45		
1025	0.25	0.45		
1045	0.45	0.80		
1080	0.80	0.80		

*Nominal composition – unless marked otherwise, the listed values are the target compositions. Each grade has tolerances that allow for some variation in the composition.

The second most common steel class is low alloy steel which is useful for its high strength potential and high hardenability. As well as the elements included in carbon steel, they can also contain molybdenum, chromium and/or nickel additions. **Table B.2** lists five of the most frequently used low alloy steels. These grades are most often used in the quenched and tempered condition.

Table B.2: Common Low Alloy Steel Grades

SAE Grade	Nominal Composition (%)							
	C	Mn	Si	S	P	Mo	Cr	Ni
4040	0.40	0.80	0.25	0.040 max	0.030 max	0.25	-	-
4130	0.30	0.50				0.20	0.95	-
4140	0.40	0.85				0.20	0.95	-
4340	0.40	0.70				0.25	0.80	1.80
8620	0.20	0.80				0.20	0.50	0.55

B.2 High Strength Low Alloy (HSLA) Steel

Whereas conventional steels use carbon for strength, 'high strength low alloy' (HSLA) steels use a different strategy – fine ferrite. Although ferrite is a weak phase, when the ferrite 'grains' get incredibly small, there is a substantial strength benefit. HSLA steels have small additions of titanium, vanadium, aluminum and/or niobium that help in the formation of ultra-fine grained ferrite microstructures. Combined with high manganese, these ultra-fine structures provide higher strengths than conventional low carbon steels. **Figure B.1** compares the microstructures of a low carbon steel with an HSLA steel. Since their carbon contents are low, HSLA steels also have good ductility, formability and are easy to weld.

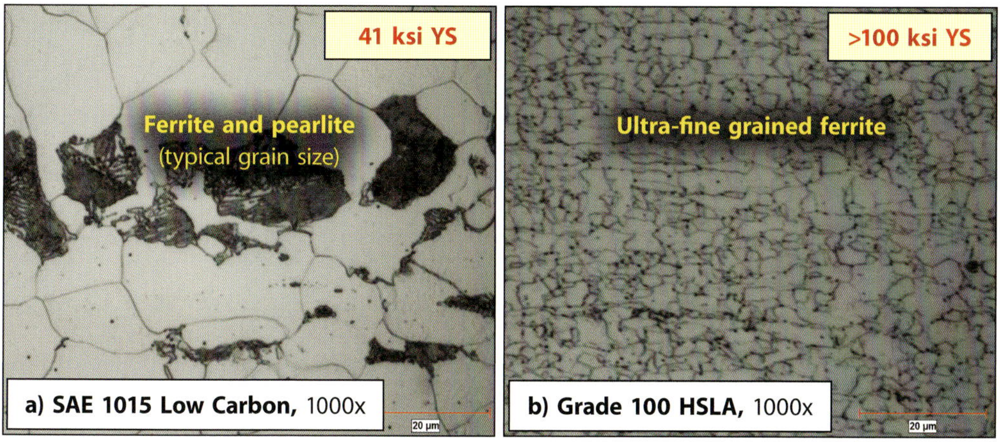

Figure B.1: Micrographs comparing the microstructures of (a) 0.15% C low carbon steel and (b) HSLA Grade 100 steel. The minute alloying of titanium, vanadium, aluminum and/or niobium forms ultra-fine ferrite grains. The fine grain size provides improved strength without sacrificing weldability.

HSLA grades are usually named by their minimum yield strength requirement. For example, HSLA Grades 65, 80 and 100 must have yield strengths greater than 65, 80 and 100 ksi respectively (**Table B.3**). With only a few boundaries, each steelmaker is given the freedom to choose their own recipes provided the steel is low carbon (weldable) and meets its minimum strength requirements.

Table B.3: ASTM A1011 HSLA Steels

Grade	YS (ksi)	UTS (ksi)	Composition (%)						
			C	Mn	S	P	Ti	V	Nb
Gr. 65	>65	>75		<1.50	<0.040	<0.04			
Gr. 80	>80	>90	<0.15	<1.65	<0.025	<0.02	0.05 minimum		
Gr. 100	>100	>110		<2.00	<0.025	<0.02			

Make no mistake, HSLA steels are not nearly as strong as quenched and tempered hardenable steels. Yet they are (1) much stronger than low carbon steel and (2) far easier to weld and form than hardenable steel. This combination makes HSLA steel ideal for a variety of applications that need good strength, formability and weldability (**Figure B.2**). For example, HSLA steels' higher strengths over carbon steel are used to reduce the weight of cars which increases their fuel efficiency – the stronger the steel, the less weight in steel that is needed.

Figure B.2: HSLA steels are used for applications where their combination of good strength, formability and weldability is beneficial. Common applications include car bodies, buildings and pipelines.

B.3 Resulfurized Steel

For over 99% of steel, sulfur is considered an impurity and efforts are made during steelmaking to minimize it. Yet there is one steel class that intentionally adds sulfur to maximize its machinability. The sulfur separates from the steel and forms a high density of 'inclusions' within the steel. This class of steel is perhaps the easiest to machine but it comes at the sacrifice of strength in particular orientations. **Table B.4** lists the composition of SAE 12L14, the most common resulfurized steel grade, that has over ten times the sulfur content of conventional steels. The 'L' in its designation indicates lead alloying was also included. **Figure B.3** provides an example of its microstructure, illustrating its high density of sulfide inclusions.

Table B.4: Resulfurized Steel

SAE Grade	Nominal Composition (%)				
	C	Mn	S	P	Pb (lead)
12L14	0.15 max	1.00	0.30	0.09 max	0.25

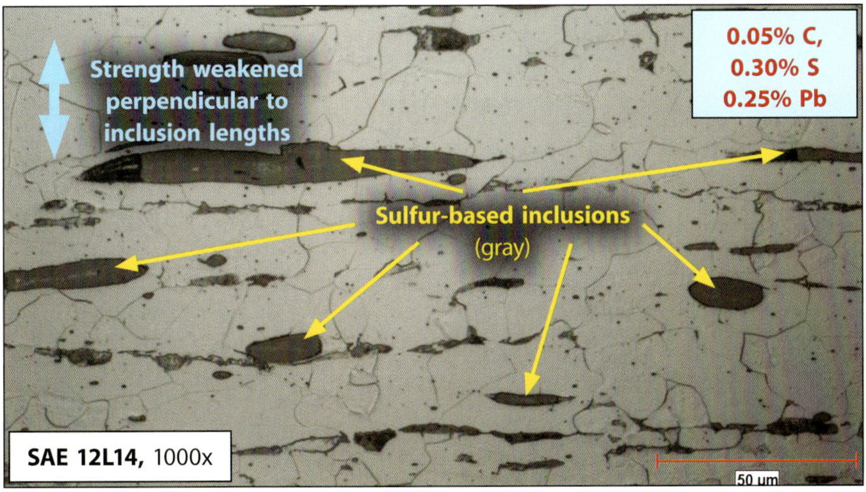

Figure B.3: The addition of sulfur to SAE 12L14 forms a high density of 'inclusions' (gray phase). Where inclusions in most steels are unwanted and efforts are made to minimize them, for this steel class they are intentionally created to improve machinability. This improved machinability comes at a strength loss perpendicular to the inclusion orientation.

B.4 Tool Steel

Tool steels are designed to optimize wear resistance and strength. They are used for applications such as cutting, stamping and shaping where the tooling must be more durable than the materials they handle. Tool steels can have high carbon contents, lots of other alloying and are always used in the tempered martensite condition. Some tool steels have sufficiently high alloying and hardenability that they form martensite even when air cooled. **Table B.5** lists a few common tool steel grades. Although governed by SAE, tool steels use a different code series than the four-digit designation system of the other classes previously introduced.

Table B.5: Common Tool Steels

SAE Grade	Nominal Composition (%)								
	C	Mn	Si	S	P	Mo	Cr	Ni	V
A2	1.00	1.00	0.50	<0.03	<0.03	1.15	5.12	0.3	0.35
D2	1.50	0.60	0.60	<0.03	<0.03	0.95	12.0	-	1.10
H13	0.38	0.35	1.00	<0.03	<0.03	1.50	5.12	0.3	1.00

Some tool steels have such high carbon and alloying that they contain large carbides which do not dissolve during heat treating. **Figure B.4** illustrates how the microstructure of SAE D2 tool steel has large 'primary' carbides that do not dissolve during heat treatment plus 'secondary' carbides that form the usual way during tempering. This carbide combination makes SAE D2 very hard and wear resistant at the expense of almost all ductility.

The primary benefit of tool steels is that they do not wear or dull quickly. On the flip side, they have almost no ductility and are difficult to machine once hardened. That is why tool steels are often purchased in the annealed condition, machined to shape and then heat treated to their final condition. After hardening to martensite and light tempering, only finishing touches are usually applied – such as touching up surfaces or sharpening edges by grinding.

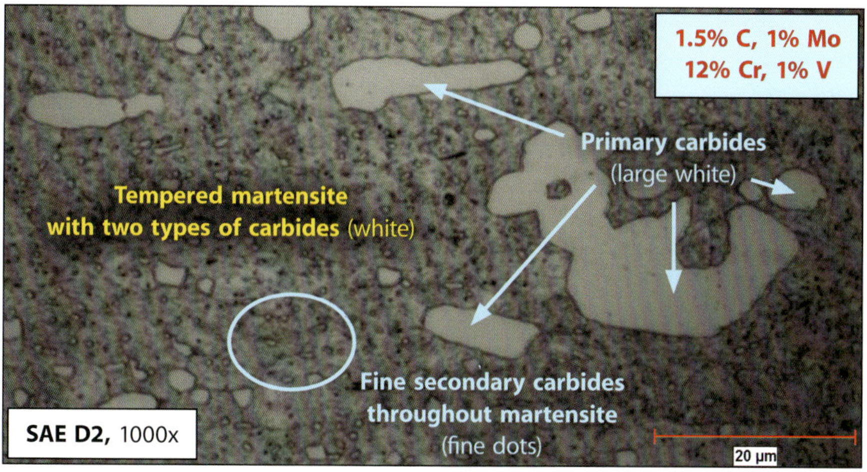

Figure B.4 D2 tool steel is saturated with carbides. Due to its 1.5% carbon plus ample molybdenum and chromium alloying, not all of the carbides dissolve into the matrix during heat treatment. These large, undissolved carbides are called 'primary' carbides while the fine carbides that form during tempering are called 'secondary' carbides. This combination of carbides provides high strength and wear resistance.

Tool steels are used for applications that must resist wear from repeated contact against other metals or materials in order to perform properly. **Figure B.5** illustrates examples of components made of tool steel that must remain sharp, dimensionally accurate or maintain a smooth surface to continue to function well.

Figure B.5: Tool steels are used because of their resistance to wear and dulling. This makes them useful for metal-on-metal applications such as tooling used to shape and cut conventional steel. Gun barrels are also made of tool steel to resist wear from metal bullets; needed to maintain their accuracy. Another common application is molds for high volume plastic injection molding.

B.5 Bearing Steel

SAE 52100 and M50 steels are tool steels used almost exclusively for roller bearings. SAE 52100 is the most common bearing alloy. Its high carbon and chromium contents give it incredibly high strength under compressive loading and excellent wear resistance – critical properties needed for bearings (**Table B.6**). **Figure B.6** displays the common microstructure of SAE 52100 which comprises of fine primary and ultra-fine secondary carbides within a lightly tempered martensite matrix. SAE M50 has even higher chromium alloying plus has molybdenum and vanadium making it even more durable (and more costly). Bearings should never be welded.

Table B.6: Common Bearing Steels

SAE Grade	Nominal Composition (%)							
	C	Mn	Si	S	P	Mo	Cr	V
52100	1.00	0.35	0.25	<0.015	<0.015	-	1.50	-
M50	0.83	0.30	0.40	<0.030	<0.030	4.35	4.12	1.05

Figure B.6: SAE 52100 is the most common alloy used for roller bearings. Its high, 1% carbon content forms both fine primary carbides and ultra-fine secondary carbides throughout its structure. SAE 52100 steel is always quenched and lightly tempered.

B.6 Stainless Steel

Chromium alloying BELOW 11% is added to increase steel strength, hardenability and sometimes wear resistance. Yet chromium alloying ABOVE 11% introduces a different benefit – it makes the steel resistant to corrosion. Above this chromium threshold, steel forms a protective chromium oxide (Cr_2O_3) on its surface that prevents corrosion. Since they do not rust like other steels, when first discovered they were called 'stainless steel'. That name is still used today.

As discussed in Chapter 10, there are a variety of stainless steel classes. Each class has different alloying and microstructural strategies to optimize different properties. All require at least 11% chromium to have the self-protecting qualities needed to be considered stainless. Most classes contain additional alloying to optimize or balance properties.

Table B.7 lists the most common grades for the five classes of stainless steel. As with all other steels, the composition is only half of the equation with the microstructure being the other half. **Figure B.7** illustrates the common microstructures for these five classes. Austenitic, martensitic and ferritic stainless steels are named after their predominant microstructure – the benefits of each having already been discussed in Chapter 10.

Table B.7: Common Stainless Steel Grades

Grade	Nominal Composition (%)							
	C	Mn	Si	S	P	Cr	Ni	Other
Austenitic Stainless Steel (Good corrosion resistance, easy to weld/manufacture, moderate strength)								
304L	<0.03	<2.00	<0.75	<0.030	<0.045	19.0	10.0	-
316L	<0.03	<2.00	<0.75	<0.030	<0.045	17.0	12.0	2.5 Mo
Martensitic Stainless Steel (High strength, moderate corrosion resistance, difficult to weld)								
410	<0.15	<1.00	<1.00	<0.030	<0.040	12.5	-	-
440C	1.10	<1.00	<1.00	<0.030	<0.040	15.0	-	<0.75 Mo
Ferritic Stainless Steel (Cheapest, low-to-moderate corrosion resistance, easy to weld/manufacture, weakest)								
409	<0.08	<1.00	<1.00	<0.30	<0.045	11.2	-	Trace Ti
Duplex Stainless Steel (Good corrosion resistance, improved chloride resistance, weldable)								
2205	<0.03	<2.00	<1.00	<0.020	<0.030	22.0	5.0	3.2 Mo
Precipitation Hardened Stainless Steel (High strength, good corrosion resistance, weldable)								
15-5	<0.07	<1.00	<1.00	<0.030	<0.040	14.75	4.50	3.5 Cu
17-4	<0.07	<1.00	<1.00	<0.030	<0.040	16.25	4.00	4.0 Cu

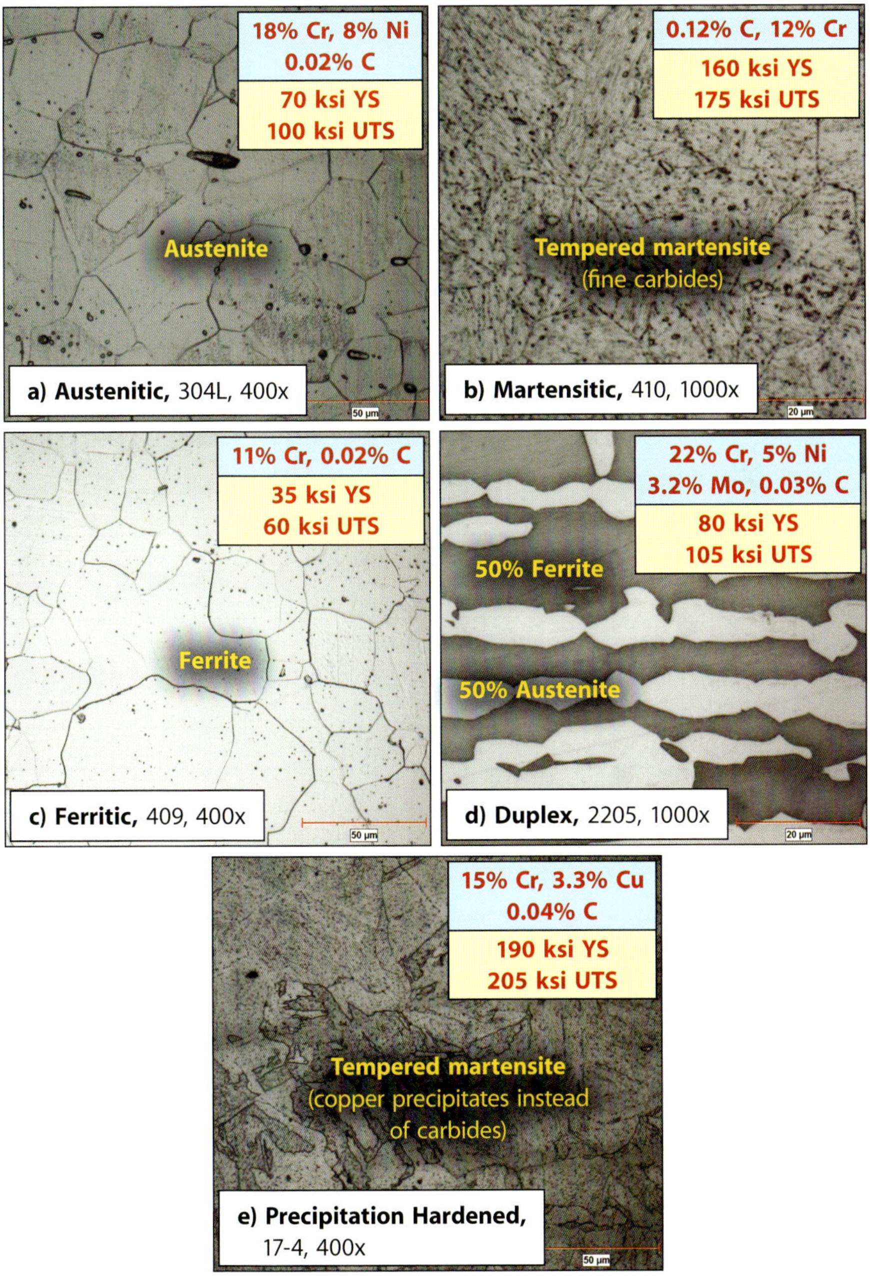

Figure B.7: Each of the stainless steel classes have different microstructures. Their combinations of alloying and microstructure provide differing balances of corrosion resistance, strength, cost and the ease/difficulty with which components can be manufactured.

One of the weaknesses of austenitic stainless steels, the most common class, is that they suffer corrosion cracking when exposed to hot water containing chlorides. 'Duplex' stainless steels were designed with microstructures of half ferrite and half austenite to make them more resistant to water-based environments containing chlorides. Duplex stainless steels have many of the same benefits of austenitic stainless steels but also have the advantage of its ferrite phase inhibiting against chloride corrosion cracks.

The last class is 'precipitation hardened' stainless steel. Steels within this class comprise of tempered martensite microstructures formed by quenching and tempering. The difference is that instead of the fine precipitates being carbides, the precipitates are copper or aluminum-based. This provides high strength without the sacrifices high carbon usually brings. This class balances high strength, good corrosion resistance, remains weldable and is relatively easy to process during manufacture. It is also quite expensive and difficult to find for purchase.

B.7 Summary of Common Steel Grades

1. Carbon steel is by far the most produced steel class in the world. This class is named after its primary strengthening element – carbon.

   a. Further divided into low, medium and high carbon subclasses.

2. The second most common steel class is low alloy steel. These grades include the addition of molybdenum, chromium and/or nickel alloying which increases their strength and hardenability.

3. High strength low alloy (HSLA) steels are low carbon and use ultra-fine ferrite for strengthening.

   a. Small additions of titanium, vanadium, aluminum and/or niobium help refine their grain sizes.

   b. Notably stronger than other low carbon steels and are easy to weld.

4. Resulfurized steels intentionally have sulfur added (which is normally considered an impurity). This improves machinability by sacrificing strength in particular orientations.

5. Tool steels and bearing steels are optimized for high strength and to resist wear. They have high carbon contents and copious amounts of other alloying.

6. Stainless steels contain at least 11% chromium which dramatically improves their corrosion resistance. Most stainless steel classes include additional alloying and use varying microstructural strategies to optimize different properties.

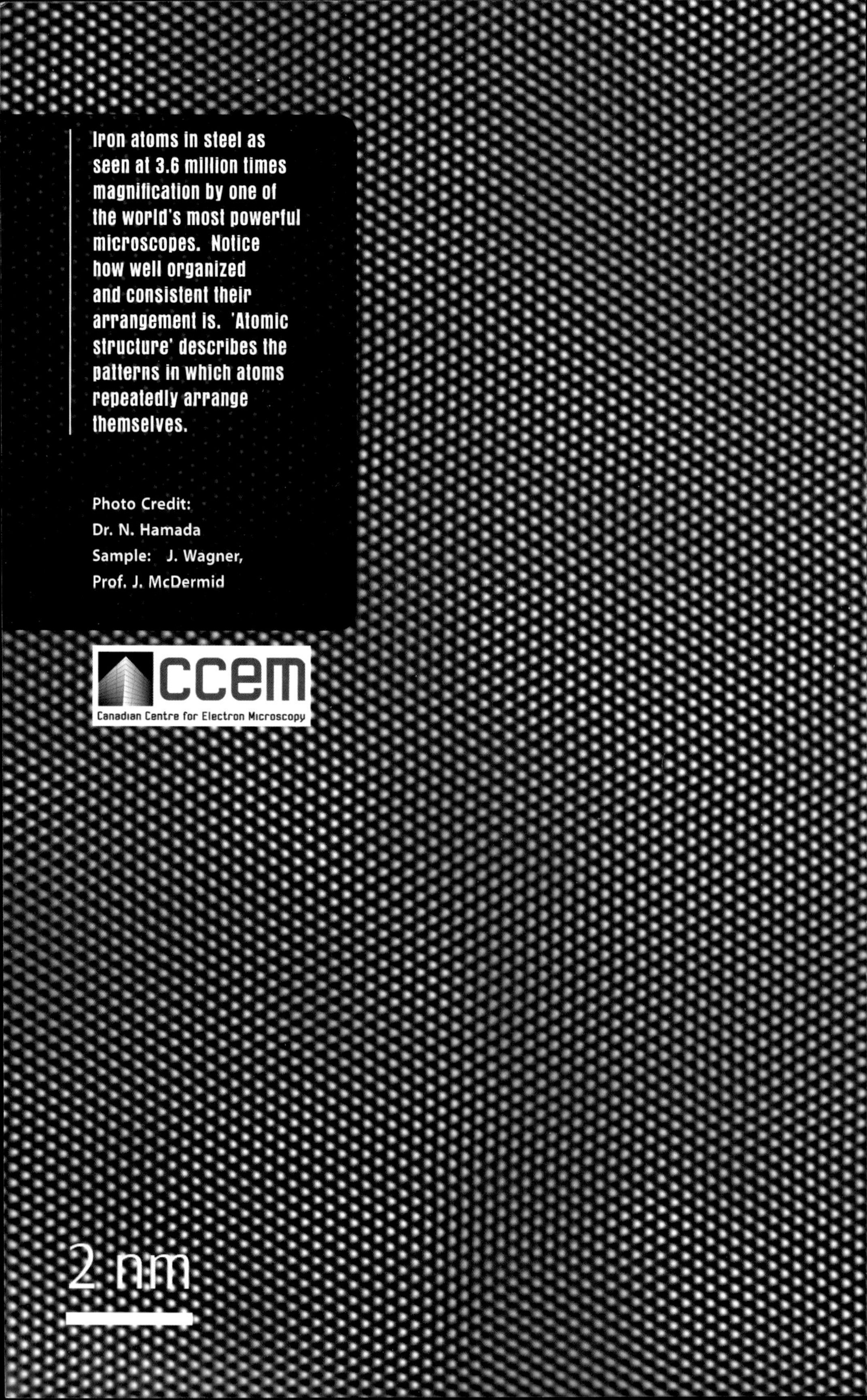

Iron atoms in steel as seen at 3.6 million times magnification by one of the world's most powerful microscopes. Notice how well organized and consistent their arrangement is. 'Atomic structure' describes the patterns in which atoms repeatedly arrange themselves.

Photo Credit:
Dr. N. Hamada
Sample: J. Wagner,
Prof. J. McDermid

ccem
Canadian Centre for Electron Microscopy

APPENDIX C: ATOMIC STRUCTURES, PHASE DIAGRAMS AND CARBON EQUIVALENT

The microstructures observed under conventional microscopes can be used to deduce the atomic structure within the steel. For those interested, the following discusses the atomic arrangements of steel and their association with microstructure. Included is an introduction to 'phase diagrams' which are used to summarize the microstructures formed at different temperatures and compositions.

C.1 Atomic Arrangement

When beginning to discuss steel at the atomic level, we must first consider the iron and carbon atoms. With most conventional steels comprising of over 98% iron, the relevance of the iron atom is obvious. On the other hand, carbon atoms make up only a tiny fraction of a steel's composition. Its substantial influence on steel is because of how the carbon atoms interact with the iron atoms. Following the carbon atoms explains quite a bit about steel behavior.

Throughout the temperature range we use steel, iron atoms are arranged in patterns called 'body centered cubic' (BCC). **Figure C.1** shows how each iron atom is surrounded by eight other iron atoms in a cubic arrangement. Or perhaps easier to visualize, an iron atom sits within the center of a cube of iron atoms. This BCC arrangement is repeated over and over again throughout the steel, creating a well-organized atomic structure.

While iron atoms are in the BCC arrangement, the gaps between them are quite small. Although carbon atoms are tiny compared to iron, the spaces between the iron atoms are too small for the carbon to fit within.

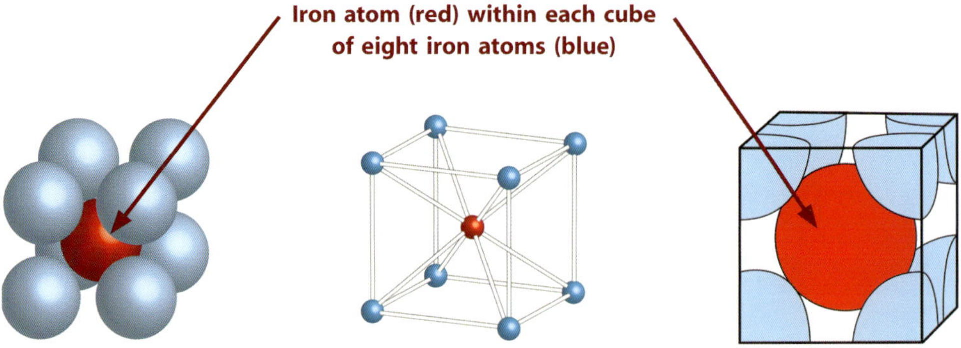

Figure C.1: The atomic structure of the steel we touch and use is body centered cubic (BCC). Each iron atom (red) is surrounded by eight other iron atoms (blue).

As a result, carbon cannot coexist inside a BCC iron matrix. Instead, carbon exists in the separate form of carbides outside the BCC matrix. Carbide (Fe_3C) is an independent, discretely different phase from the BCC iron matrix that surrounds it.

How this relates to steel microstructure is that ferrite is effectively carbon and carbide-free BCC. Ferrite has neither carbon atoms in its BCC matrix nor does it have carbides scattered within. On the other hand, pearlite has a carbon-free BCC matrix yet is intermixed with carbide plates (Fe_3C). These carbide plates create the distinctive bands that make pearlite recognizable under conventional microscopes (**Figure C.2**).

In other words, ferrite contains essentially no carbon – its BCC iron atom matrix is carbon-free and it does not have carbides. Pearlite also has a carbon-free BCC matrix (no carbon between the iron atoms) but is intermingled with lamellar Fe_3C carbide plates.

Why steel microstructures are reset at red-hot temperatures is because the iron atomic structure begins to change to 'face centered cubic' (FCC) above 1340°F (>727°C). FCC is a more complex arrangement than BCC. In the FCC arrangement, an iron atom is situated at each of the six faces of a cubic structure (**Figure C.3**). FCC in conventional steel is only stable at red-hot temperatures and, upon cooling down, the steel will revert back to BCC. The microstructure name corresponding to steel in the FCC arrangement is 'austenite'.

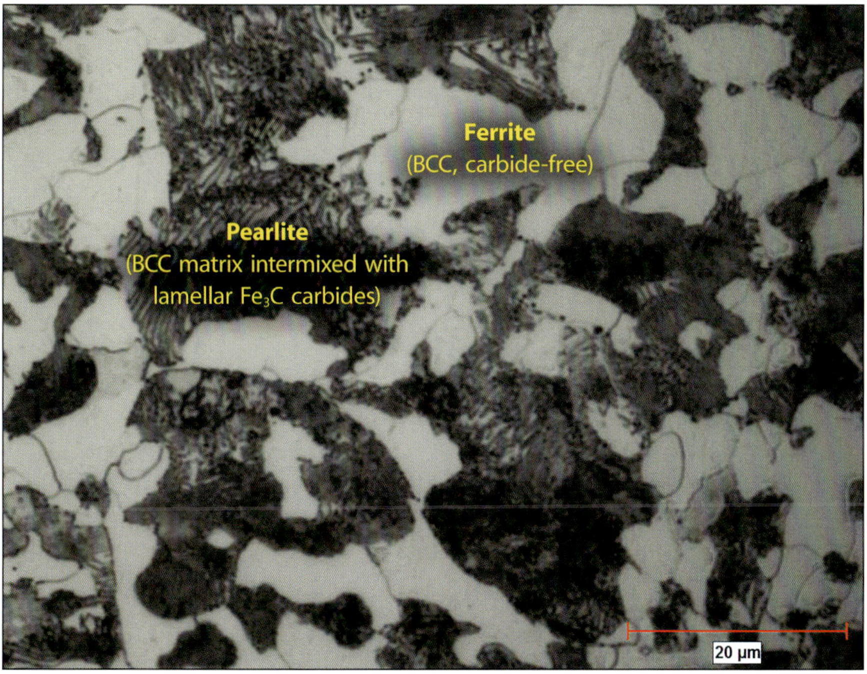

Figure C.2: The terms used to describe the microstructure are simplifications of the atomic arrangements. Ferrite is carbon-free BCC without any carbides. Pearlite is a carbon-free BCC matrix intermixed with thin carbide plates. These carbide plates make pearlite distinguishable under a microscope.

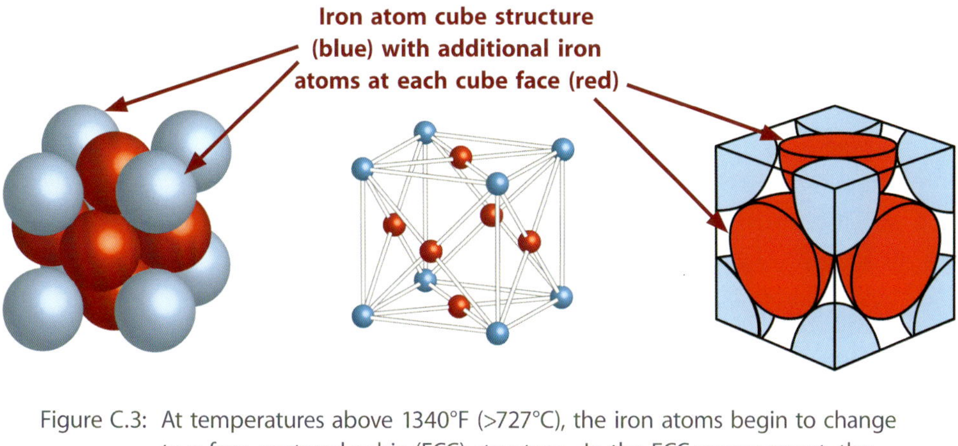

Figure C.3: At temperatures above 1340°F (>727°C), the iron atoms begin to change to a face centered cubic (FCC) structure. In the FCC arrangement, the gaps between the iron atoms are large enough for carbon atoms to fit within. When a steel is in the FCC arrangement, its microstructure is referred to as 'austenite'.

The iron atomic arrangement changing to FCC at high temperatures has two useful effects. The first is that some of the gaps between the FCC lattice are large enough that the small carbon atoms can fit within. While in the FCC arrangement, preexisting carbides dissolve and the carbon atoms become dispersed between the iron atoms. This dissolution is key to resetting the microstructure. When the steel cools down and the matrix switches back to BCC, the carbon atoms will no longer comfortably fit between the iron atoms. What structure forms upon cooling depends largely on how much carbon had dissolved into the matrix and how much time the carbon has to escape.

The second useful effect is that the FCC austenite structure with its lack of carbides is weak, soft and far more ductile than the BCC arrangement. The low strength of austenite is what makes steel easy to forge and shape at red-hot temperatures.

C.2 Phase Diagrams

To properly reset the microstructure during heat treating, the steel microstructure needs to transform entirely into austenite (FCC). Although all conventional steels begin to transform from BCC to FCC at 1340°F (727°C), the temperature at which the steel becomes completely FCC depends upon its alloying – its carbon content in particular.

There is ample literature providing heat treatment temperatures and procedures for each grade of steel. It is recommended that when choosing the heat treatment parameters for a particular grade that such literature is used. Yet we can use the iron-carbon phase diagram to explain the theory behind how the heat treatment temperatures have been established. Plus, these diagrams illustrate the basic principles of how carbon and temperature affect the microstructure.

Figure C.4 illustrates the portion of the iron-carbon phase diagram most relevant to the use of steel (most steels have less than 0.8% carbon). Phase diagrams summarize what atomic structures and phases are stable at different temperatures and compositions. If the heating and cooling rates are slow, it predicts what microstructure is formed. In some areas

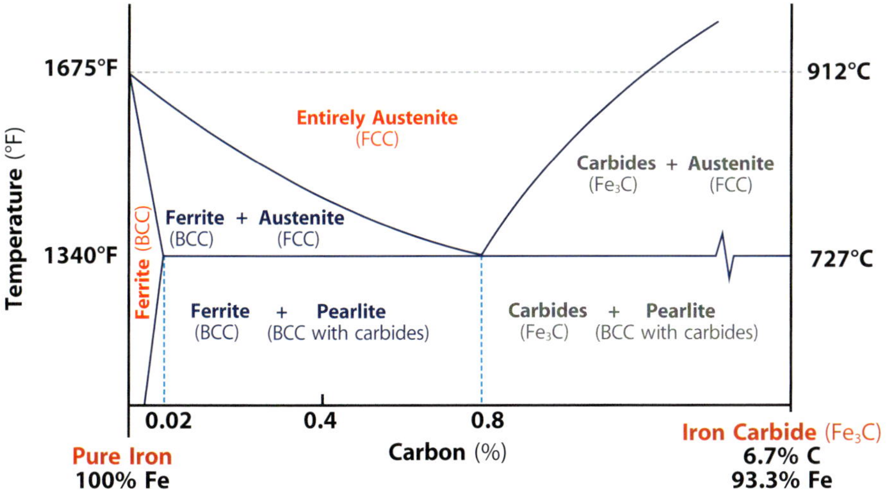

Figure C.4: The iron-carbon phase diagram illustrates the effects that temperature and carbon content have upon the atomic arrangements and microstructure of steel. Most steel has less than 0.8% carbon.

of the diagram, only single phases of BCC, FCC or iron carbide (Fe_3C) are stable. Mixed phases are expected in the areas between.

One use of the phase diagram is that it provides the minimum temperature needed to fully 'austenitize' steel and transform it entirely into FCC. The diagram illustrates how that temperature decreases with increasing carbon content up until 0.8% carbon. To provide a safety margin and ensure full transformation, heat treatments such as normalizing are completed 50-100°F above the theoretical austenite transformation temperature (**Figure C.5**).

For example, although all three bolts had been normalized in Chapter 2 at the same temperature of 1650°F (~900°C), the higher carbon steel bolts could have been heat treated at lower temperatures. **Figure C.5** uses the phase diagram to demonstrate how the Grade 5 (0.40% C) bolt could have been normalized at 1550°F (843°C). This is one hundred degrees below that needed to normalize the Grade 2 (0.15% C) bolt.

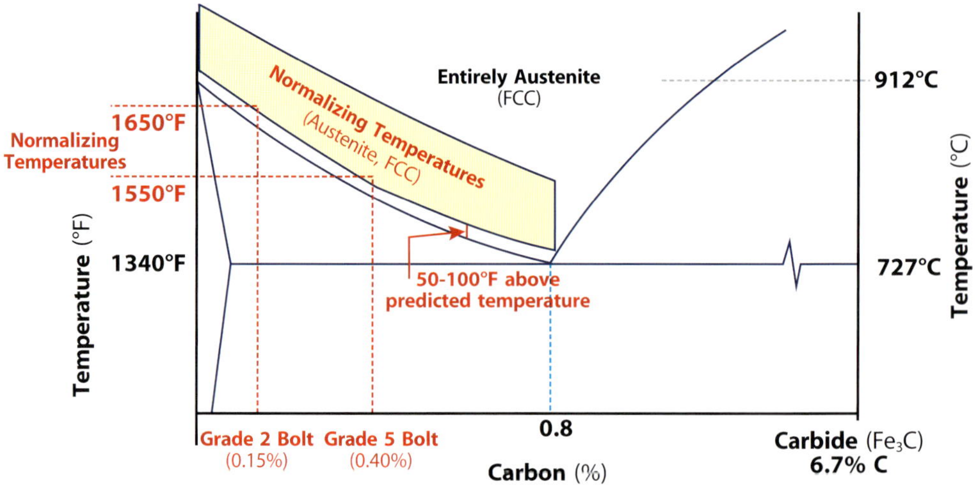

Figure C.5: The iron-carbon phase diagram summarizes what minimum temperature is needed to austenitize steel based upon its carbon content, transforming it into FCC and resetting its microstructure. Steel with higher carbon fully austenitize at lower temperatures.

Phase diagrams can also be used to help explain the formation of the phases we see under a microscope. When steel is slowly air cooled from austenitic temperatures, the diagram explains what is happening within. Below 0.8% carbon, ferrite is formed upon cooling from austenitic temperatures down to 1340°F (727°C). The higher the carbon content, the less ferrite that will form – 0.8% carbon has zero ferrite. For a given carbon content, the relative distance between 0 and 0.8% carbon can be used to approximate how much ferrite will be formed. For example, **Figure C.6** illustrates why half the Grade 5 (0.4% C) bolt microstructure comprised of ferrite.

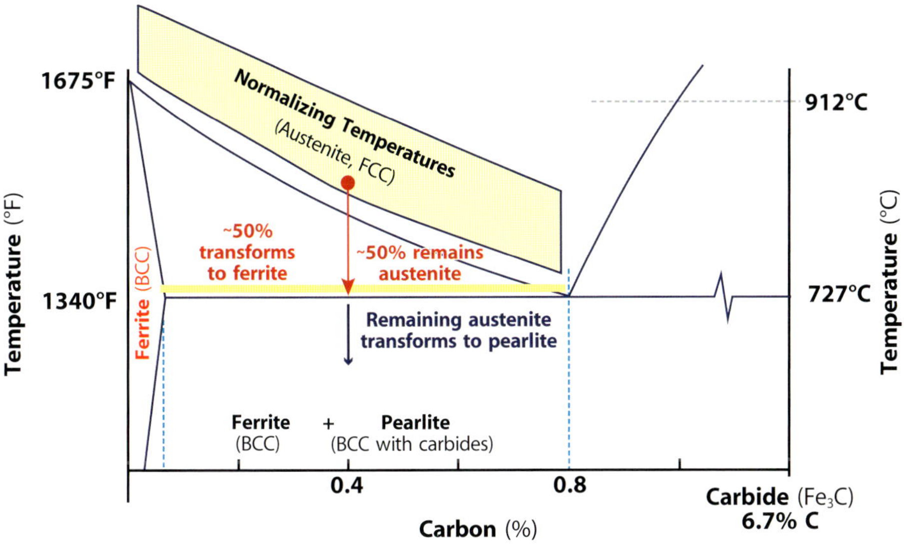

Figure C.6: Phase diagrams explain the microstructures that form during slow cooling. When cooled from austenitic temperatures, ferrite forms first. The amount of ferrite that forms before reaching 1340°F (727°C) depends on the steel's carbon content. In the case of a Grade 5 (0.4% C) bolt, half its structure would have transformed to ferrite while the rest remained austenite. Upon crossing below 1340°F (727°C), the remaining austenite transforms into pearlite – a BCC matrix intermixed with carbide plates.

Upon reaching 1340°F (727°C), all remaining austenite (FCC) now transforms into pearlite. The steel matrix becomes BCC and the carbon atoms are rejected from the iron matrix to form the plate carbides distinctive to pearlite.

Learning to read phase diagrams can be a powerful tool for understanding steel behavior. Plus, phase diagrams can be helpful for the study of other metals. For those who are interested in studying materials in greater detail, consider taking the time to learn more about phase diagrams.

C.3 Carbon Equivalent

Nothing affects the behavior of steel more than carbon. When contemplating how carbon steel or lightly alloyed steel behaves, considering the effects of carbon is usually enough to understand how the steel will respond to processing. However, when significant amounts of other alloying are included, their combined contributions to steel behavior need to be considered. To amalgamate the effects of all relevant alloying within a steel, 'carbon equivalent' (CE) is used. Carbon equivalent is commonly calculated by:

$$Carbon\ Equivalent\ (CE) = C + \frac{Mn}{6} + \frac{Cr + Mo + V}{5} + \frac{Cu + Ni}{15}$$

Carbon equivalent (CE) provides a more accurate representation of how steel will behave according to the phase diagram – the steel is charted using its CE value along the carbon scale. It is also useful in defining whether alloyed steel is hardenable. As a rule-of-thumb, carbon steel typically becomes hardenable with 0.30% carbon (this takes into account that manganese will also be present in carbon steel). When using carbon equivalent and factoring in other alloying, the transition to hardenability begins to occur at about 0.35% CE. Carbon equivalent is used in assessing the appropriate welding methods and whether pre/post-weld heating is required.

C.4 Holding Times During Heat Treating

The changes that occur as described by the phase diagram happen rather quickly. Not as quickly as being water quenched but if the temperature is adjusted within a furnace or by air cooling, the microstructural changes occur within a time frame of seconds to several minutes.

When heat treating, the amount of time a part is heated must be considered to ensure the part's microstructure is completely austenitized. 'Hold time' is the term used to refer to how long a part is held at temperature

– the clock starting once the furnace reaches the intended temperature. Part size needs to be considered as it will take longer for larger parts to reach temperature compared to smaller parts. Yet that difference in time is not quite as long as most people might think.

Standards exist such as AMS 2759 that include minimum hold times based upon the part size. Even though these listed hold times are more than adequate, often industry will unnecessarily hold for even longer durations – sometimes hours longer. The most common rules-of-thumb used by industry include (a) one hour for every inch of steel thickness or (b) one hour for the first inch of thickness plus thirty minutes for each inch afterward. These are gross overestimations and a legacy of bad habits established almost a century ago. As the need to combat climate change intensifies, opportunities remain to optimize hold times. Optimization could reduce heat treatment costs, increase furnace throughput and reduce emissions – all without compromising the part quality.

C.5 Summary of Atomic Structures, Phase Diagrams and Carbon Equivalent

1. Within the temperature range we use steel, the iron atoms are arranged in an atomic structure called 'body centered cubic' (BCC).
    a. The gaps between the iron atoms are too small for carbon atoms to fit into.
    b. Ferrite is essentially carbon-free BCC without carbides.
    c. Pearlite is carbon-free BCC intermixed with the separate phase of lamellar carbides (Fe_3C).
2. At red-hot temperatures, beginning at 1340°F (727°C), the atomic structure begins to transform into face centered cubic (FCC).
    a. The gaps between the iron atoms are large enough for carbon to fit into. Carbides dissolve and carbon enters

the FCC matrix. This is what resets the microstructure during heat treatment.

b. The microstructural term for the FCC phase is 'austenite'.

c. To successfully reset the microstructure during heat treatment, the steel must be fully transformed to austenite. The temperature needed to fully austenitize steel depends upon its composition.

3. Phase diagrams illustrate the phases and atomic structures that are expected at different compositions and temperatures.

   a. The iron-carbon phase diagram explains the microstructures of steel that will form upon heating and slow cooling.

   b. Microstructural changes take between seconds to several minutes to occur.

4. Carbon equivalent (CE) considers the combined effects of carbon and other alloying on steel behavior.

   a. CE better represents well-alloyed steels when using the phase diagram.

   b. Steels become hardenable at about 0.35% CE.

$$Carbon\ Equivalent\ (CE) = C + \frac{Mn}{6} + \frac{Cr + Mo + V}{5} + \frac{Cu + Ni}{15}$$

Long ago, Japanese blacksmiths used a unique heat treatment for making katana swords. They did not know that it optimized the blades' microstructure – they just knew it produced excellent swords.

APPENDIX D: QUENCHING AND TTT DIAGRAMS

D.1 Atomic Structure of Martensite

As discussed in Appendix C, at the temperatures we use steel, the stable atomic arrangement is body centered cubic (BCC). When slowly cooled from austenitic temperatures, the carbon atoms have ample time to exit the FCC matrix as it transitions to BCC.

However, if hardenable steels are quenched from austenitic temperatures, the carbon atoms between the iron atoms cannot escape in time as everything quickly contracts. During the formation of martensite, the iron atoms snap into BCC yet the carbon remains trapped between the iron, stretching and distorting the BCC structure. This carbon-filled, convoluted BCC structure is strong but brittle. The carbon within the distorted matrix prevents the iron atoms from moving or sliding needed to deform; eliminating all ductility. It is easier for the carbon-filled, distorted BCC structure to break apart than to deform.

In this highly stressed state, tempering at moderate temperatures allows for the carbon to squeeze out from between the iron atoms and escape. This sudden exodus of carbon is what forms the small, evenly dispersed carbides throughout the structure. After the carbon escapes, the matrix relaxes into the proper BCC structure which allows the iron atoms to slide and move again if needed, restoring its ductility. Plus the evenly distributed carbides strengthen the steel, providing the high strength associated with tempered martensite.

D.2 TTT Diagrams

Phase diagrams are useful during normalizing and annealing when the cooling rates are slow enough that the carbon can escape the FCC matrix. Yet when steel is quenched from austenitic temperatures, the steel does not have time to behave as described by its phase diagram. Therefore, a different series of diagrams, called 'time-temperature-transformation' (TTT) diagrams, are used to predict what microstructures will form during quenching. **Figure D.1** compares the iron-carbon phase diagram with a TTT diagram.

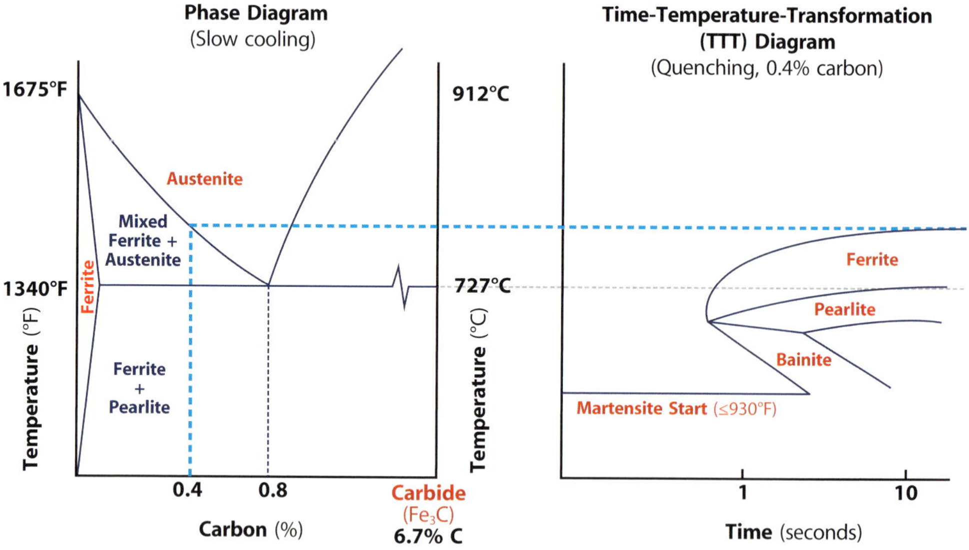

Figure D.1: (Left) While phase diagrams summarize the phases that are stable during slow temperature changes, (right) TTT diagrams illustrate what microstructures form during rapid quenching. TTT diagrams show how quickly a specific steel needs to be quenched to form martensite and what structures will form if not cooled quickly enough.

TTT diagrams are simple. They begin at the austenitizing temperature used during heat treatment and, upon quenching, the rate of cooling is plotted (**Figure D.2**). Whichever phase line the cooling plot hits, that phase will begin to form. In most cases, the goal of using TTT diagrams is to help understand what quenching rates are needed to through-harden steel into martensite. For that purpose, quenching aims to cool the steel down fast enough to intersect the 'martensite start' temperature without hitting the ferrite-pearlite 'nose' on the way down. If the cooling rate hits any other phase line before reaching the martensite start temperature, other microstructural phases will form – which could be problematic.

In other words, if the objective is to form martensite and benefit from its high strength, the TTT cooling plot must hit the 'martensite start' line uninterrupted. If steel is quenched too slowly and the plot line intersects with other phase lines, then weaker microstructures (bainite, ferrite, pearlite) will be formed.

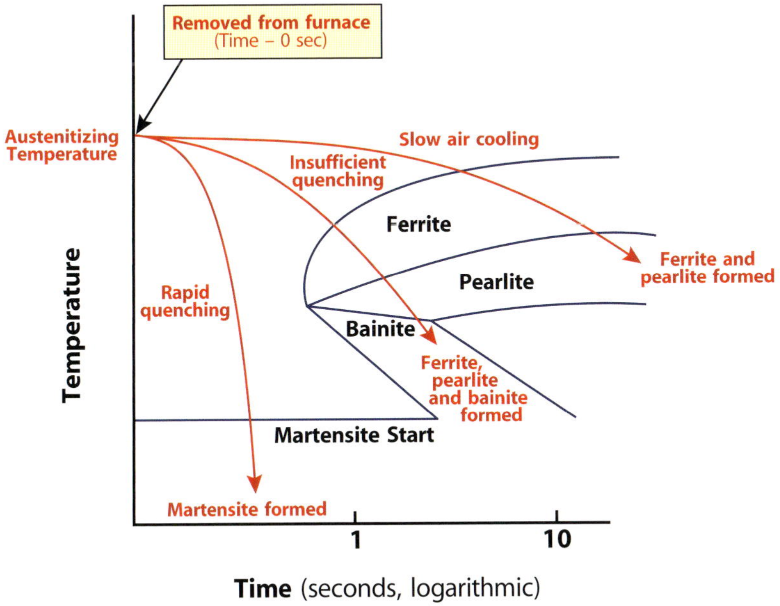

Figure D.2: A TTT diagram illustrating three different cooling rates plotted. If quenched fast enough and the cooling plot intersects the martensite start temperature without interruption, a martensitic structure will be formed. If not quenched fast enough and other phase lines are intersected, other phases will form.

Each steel grade has its own TTT diagram that describes what cooling rates are needed to form martensite. These diagrams indirectly describe why each steel grade has a maximum thickness that can be through-hardened into martensite. Remember that as the part size increases, the slower its core will quench. This effect had been demonstrated by heating and water quenching 1045 carbon steel (0.45% C) bars of various diameters. The cores of the 1-inch thick and thicker bars could not be quenched fast enough to fully through-harden them into martensite (**Figure D.3**).

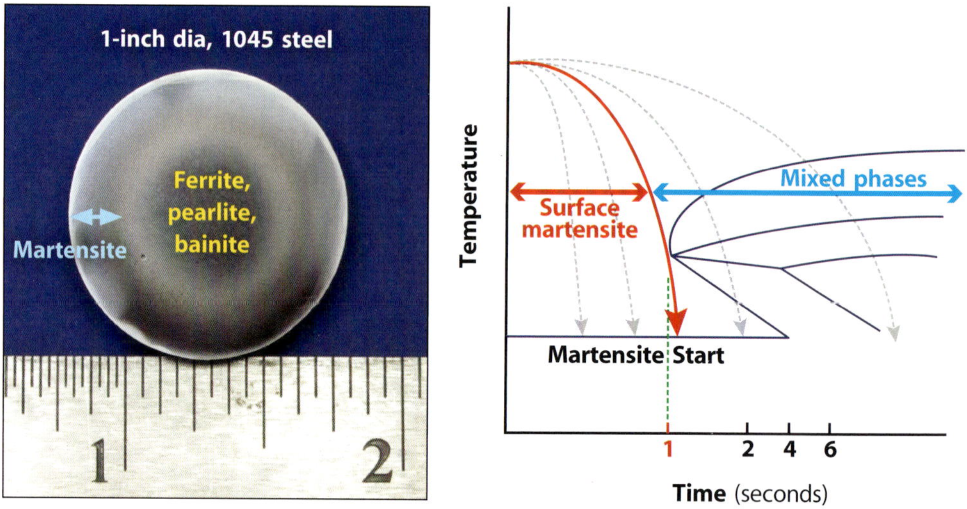

Figure D.3: Per its TTT diagram, once quenching begins, 1045 carbon steel (0.45% C) only has one second to cool down below the ferrite/pearlite nose in order to form martensite. This was why only the surface of the water quenched 1-inch diameter 1045 steel bar had formed martensite. The cooling rate at its surface was rapid enough to reach the martensite start temperature uninterrupted. Deeper within the bar, the cooling rates were slower and intersected other phase lines, forming a mixture of non-martensitic phases.

Recall that the steel thickness that can be quenched into martensite can be enlarged by increasing its alloying content. **Figure D.4** illustrates the generic effects that increasing the carbon content and other alloying has upon the TTT diagram. For low carbon, non-hardenable steels, the ferrite-pearlite nose is fully forward and no cooling rate can form martensite. Increasing the carbon, molybdenum, chromium and nickel contents shift the nose back. Steels become hardenable when the nose is back far enough that it is possible to intersect the martensite start temperature without interruption. Increasing the carbon and alloying continues to shift the nose back, allowing the steel to form martensite at slower and slower quenching rates. This is what increases the steel's 'hardenability' which in turn increases the part size that can be through-hardened into martensite.

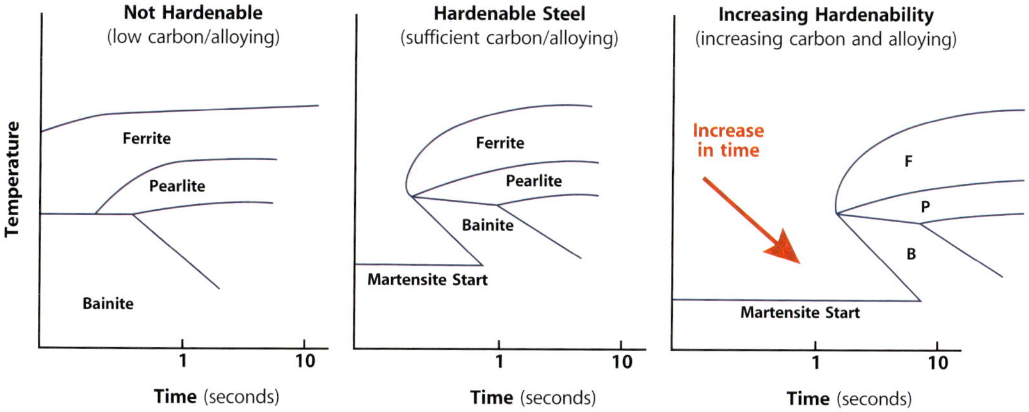

Figure D4: (Left) For steel with insufficient carbon or alloying to be hardenable, the ferrite-pearlite 'nose' intersects with the start time (0 seconds), making it essentially impossible to form martensite. (Mid, right) For hardenable steels, increasing the carbon and alloying (Mo, Cr, Ni) pushes the nose back and reduces the martensite start temperature – increasing the time the steel has to intersect the martensite start temperature. The ability to form martensite at slower quench rates is what allows for the through-hardening of thicker parts.

Figure D.5 illustrates how the additional molybdenum and chromium alloying in a 4140 low alloy steel pushed the nose back, providing more time to cool past the ferrite-pearlite nose and form martensite. This is what made it possible to through-harden much thicker sections of 4140 low alloy steel compared to 1045 carbon steel.

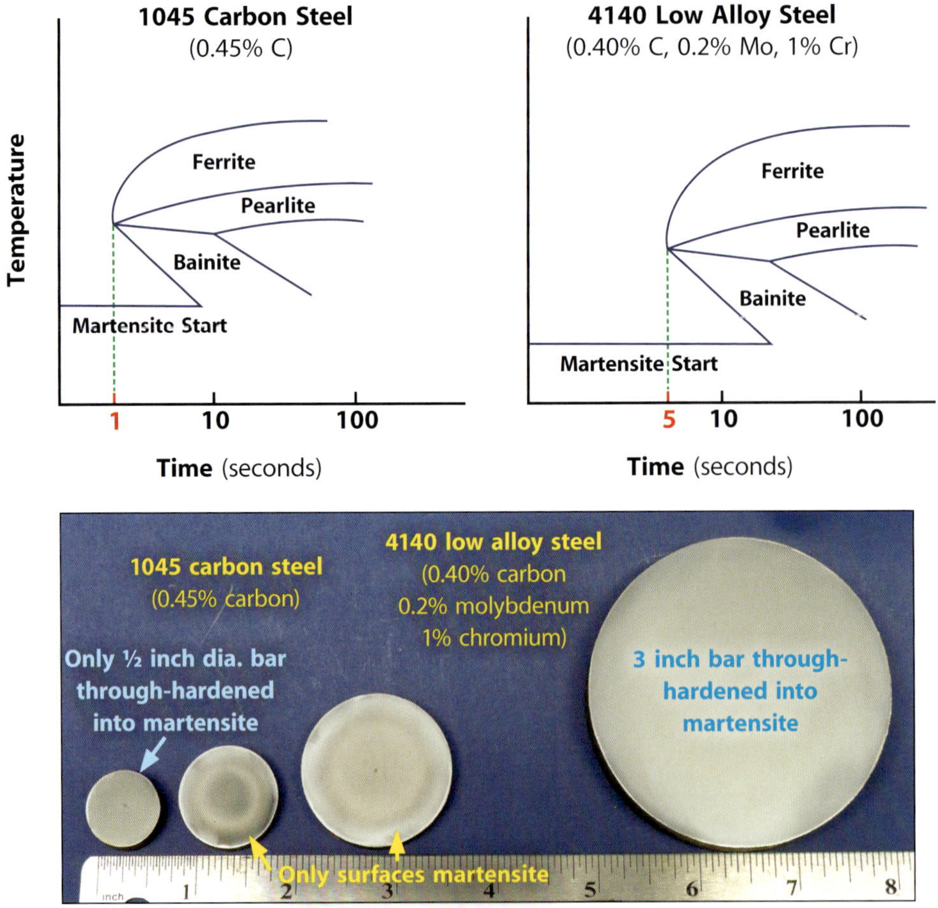

Figure D.5: Whereas 1045 carbon steel only has one second to quench past the nose to form martensite, 4140 low alloy steel has five seconds to pass its nose. This extra time is due to its molybdenum and chromium additives. Although the difference between one second and five seconds might seem trivial, it has a substantial effect upon the part size that can be through-hardened. These additional seconds are what allowed for the through-hardening of 4140 bar many times the size of the 1045 steel bar.

D.3 One Final TTT Example – The Katana Sword

The invention of steel was so monumental that an entire historical era was named after it – The Iron Age. Spanning thousands of years, the quintessential weapon of this time was the sword and societies that produced better swords benefited from its advantage. During this era and throughout the world, the best swords were arguably the weapons of the Japanese Samurai – the katana sword.

One of the design features of the katana was its clever heat treatment which can be best explained using a TTT diagram. During the final stages of swordmaking, the blade would be coated in clay, leaving only the cutting edge uncovered and exposed. The blade was then heated one last time and water quenched. The exposed cutting edge would quench to form martensite. The remainder of the blade covered in clay would be insulated from the water and would cool much slower (**Figure D.6**). This would form bainite (ideally), ferrite and pearlite throughout the rest of the blade. Depending on the smith, the sword may be tempered at a very low temperature but not hot enough to cause significant softening of the critical cutting edge.

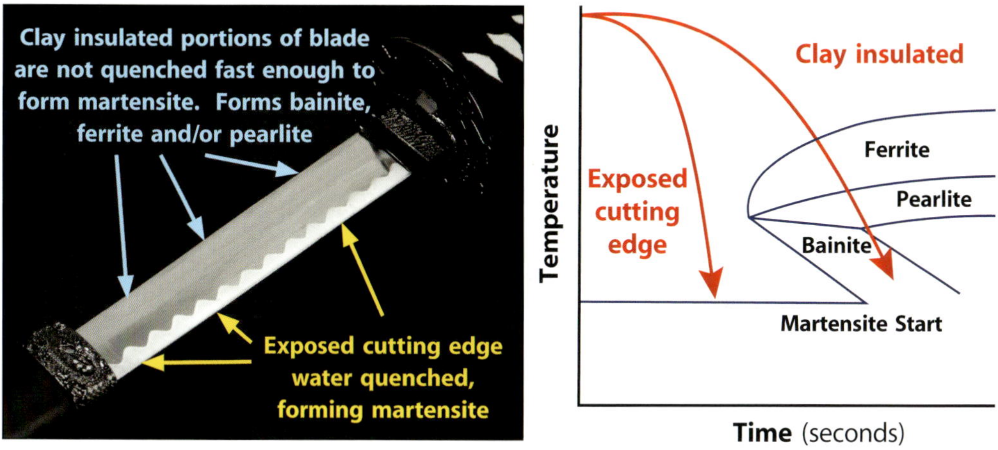

Figure D6: A TTT diagram demonstrating why the bare cutting edge of a katana blade was quenched into martensite while the remainder of the blade, covered by clay, formed other phases. The combination of the hard cutting edge and softer core created a lethal and robust blade.

The benefits of this well-designed heat treatment were similar to induction hardened parts made today – the hard martensite provided an ideal cutting edge while the remainder of the blade was softer and prevented its embrittlement. The steel robustness in this condition also allowed the blade to be thinner and lighter than most other swords, making it dangerously fast. Ultimately, because of its heat treatment and the microstructures it formed, the katana's steel properties were optimized.

Although the Japanese blacksmiths did not know why it worked, this intelligent heat treatment helped them to produce superior weapons. Its heat treatment was not the only reason why the katana was such a superb weapon. Yet since no other culture had produced swords with as effective a heat treatment, their swords were never quite as good.

Which brings us to the final idea of this book. Hopefully by now you can appreciate why the abilities to design, manufacture and even repair steel components require a practical understanding of steel behavior. Yet the advantages of learning about steel and other materials go far beyond this – it can help produce superior products. Good design is not just ensuring steel is shaped properly or simply strong enough. It includes careful thought into the best use of the materials involved. For steel parts, this includes consideration of the ideal alloying and how to optimize its properties for a particular application. If done well, like the Japanese katana, it can be the cornerstone to creating exceptional products, equipment and structures.

D.4 Summary of TTT Diagrams

1. During quenching and upon forming martensite, the sudden transition from FCC (austenite) to BCC (untempered martensite) traps the carbon atoms in the BCC matrix, distorting its structure.

    a. The trapped carbon atoms lock the iron atoms in place, preventing their ability to move and deform – making the steel strong but brittle.

    b. During tempering, the carbon is quickly ejected from the BCC matrix, relaxing the structure and returning ductility. The resulting, evenly dispersed carbides strengthen the steel.

2. Time-temperature-transformation (TTT) diagrams predict what phases will be formed during the quenching of steel.

    a. To form martensite, steel must be quenched fast enough to hit the martensite start temperature before hitting any other phase lines.

    b. Increasing the carbon and alloying content shifts the nose and phase lines such that slower quench rates can still form martensite. Martensite formation at slower quench rates makes it possible to through-harden larger parts.

ABOUT THE AUTHOR, OTHER BOOKS AND TRAINING

Although steel failures are rare, they do occur. Steel parts can fail for many different reasons ranging from excessive loading events, millions of small loading iterations, corrosion, wear or heat degradation. In order to develop effective solutions and

prevent more failures, the reasons why the part failed must be first determined. Diagnosing why steel parts failed, including whether material or manufacturing issues had played roles in their failures, is what Shane Turcott (author) does at Steel Image.

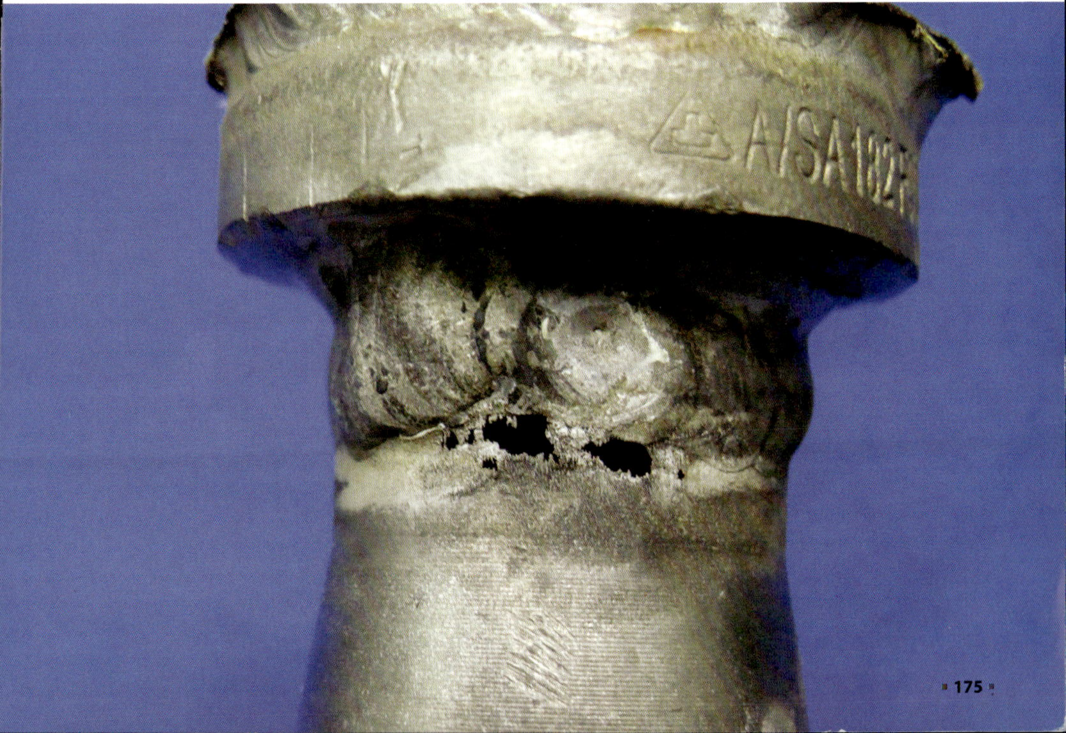

About the Author

Shane Turcott, P.Eng., received his B.Eng. and M.A.Sc. in materials engineering at McMaster University, Canada. He worked for Dofasco (steel mill), Bodycote (laboratory) and Liburdi Turbine Services (repair facility) before founding Steel Image in 2009. Throughout his career he has completed several thousand metallurgical projects and failure analyses. He is enthusiastic about exploring, explaining and increasing the interest in what he believes to be two fascinating and useful topics – steel behavior and investigating metal failures.

Other Books by S. Turcott

Decoding Mechanical Failures

Decoding Mechanical Failures introduces how to examine fractured steel parts and glean details useful for understanding why they failed. It teaches how to recognize, diagnose and interpret the fracture features of fatigue, brittle and ductile failures. It uses several well-illustrated examples to highlight relevant features of each damage mechanism. It then explains how diagnosis can be used to determine the root causes of failure. Lastly, numerous case studies to tie everything together, demonstrating how details learned from fracture features of actual investigations helped determine the causes of failure.

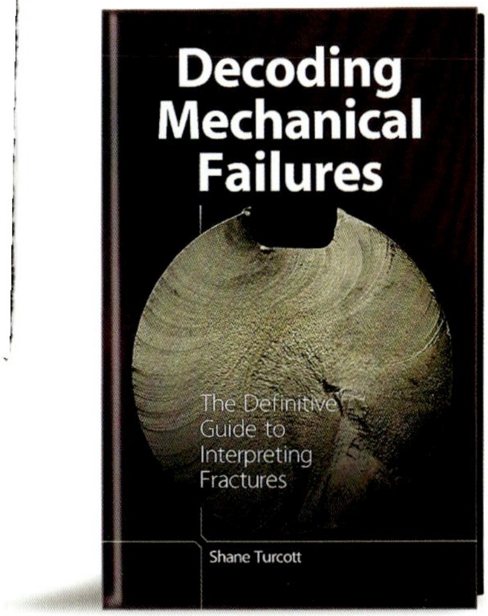

Corrosion
Case Study Courses

There are many forms of corrosion – each caused by different factors and each affecting susceptible metals in different ways.

These courses use case studies to demonstrate the effects of common corrosion modes on metals. After reviewing their practical basics, dozens of examples are used to share subtleties useful to understanding how different environments can damage metal.

STEEL IMAGE
FAIL | LEARN | SUCCEED

About Steel Image Inc.

Steel Image is a metallurgical engineering company specializing in failure analysis, laboratory testing and training. Our primary business is analyzing failed equipment suffered by the petrochemical, automotive, mining and energy industries – diagnosing the details of how equipment failed and providing information useful to preventing further failures.

Steel Image also provides training on how to examine fractured parts and begin the diagnosis process.

When a steel part breaks, the broken part itself is the greatest source of information as to why it failed.

For several common failure modes, the crack surfaces that form can be studied and interpreted to help understand why the part failed – skills that can be learned.

Introduction to Examining Fractures

Based upon the book *Decoding Mechanical Failures*, this one-day, instructor-led online course introduces how to recognize and interpret ductile, brittle and fatigue fracture features of steel parts. Participants gain experience through evaluating dozens of fractured parts, learning the subtle details useful for investigating failures.

This is a very practical course, providing skills useful for anyone dealing with equipment failures: engineers, reliability, maintenance, inspectors and metallurgists.

Learn more at www.steelimage.com/training